AF412840

CHARACTERIZATION IN COMPOUND
SEMICONDUCTOR PROCESSING

CHARACTERIZATION IN COMPOUND SEMICONDUCTOR PROCESSING

EDITORS

Yale Strausser and Gary E. McGuire

SERIES EDITORS

C. Richard Brundle and Charles A. Evans, Jr.

MOMENTUM PRESS, LLC, NEW YORK

Characterization in Compound Semiconductor Processing
Copyright © Momentum Press, LLC, 2010

First published by Butterworth-Heinemann in 1995
Copyright ©1995, by Butterworth-Heinemann, a division of Reed-Elsevier, Inc.

Reissued volume published in 2010 by
Momentum Press®, LLC
222 East 46th Street, New York, N.Y. 10017
www.momentumpress.net

ISBN-13: 978-1-60650-041-5   (hard back, case bound)
ISBN-10: 1-60650-041-4   (hard back, case bound)

ISBN-13: 978-1-60650-043-9   (e-book)
ISBN-10: 1-60650-043-0   (e-book)

DOI forthcoming

Interior Design by Scribe, Inc.

10 9 8 7 6 5 4 3 2 1

Printed in Taiwan R.O.C.

Contents

CHARACTERIZATION OF III–V THIN FILMS FOR ELECTRONIC DEVICES

III–V COMPOUND SEMICONDUCTOR FILMS FOR OPTICAL APPLICATIONS

DEEP LEVEL TRANSIENT SPECTROSCOPY: A CASE STUDY ON GaAs

APPENDIX: TECHNIQUE SUMMARIES

Preface to the Reissue of the Materials Characterization Series

The 11 volumes in the Materials Characterization Series were originally published between 1993 and 1996. They were intended to be complemented by the *Encyclopedia of Materials Characterization*, which provided a description of the analytical techniques most widely referred to in the individual volumes of the series. The individual materials characterization volumes are no longer in print, so we are reissuing them under this new imprint.

The idea of approaching materials characterization from the material user's perspective rather than the analytical expert's perspective still has great value, and though there have been advances in the materials discussed in each volume, the basic issues involved in their characterization have remained largely the same. The intent with this reissue is, first, to make the original information available once more, and then to gradually update each volume, releasing the changes as they occur by on-line subscription.

C. R. Brundle and C. A. Evans, October 2009

Preface to Series

This Materials Characterization Series attempts to address the needs of the practical materials user, with an emphasis on the newer areas of surface, interface, and thin film microcharacterization. The Series is composed of the leading volume, *Encyclopedia of Materials Characterization*, and a set of about 10 subsequent volumes concentrating on characterization of individual materials classes.

In the *Encyclopedia*, 50 brief articles (each 10 to 18 pages in length) are presented in a standard format designed for ease of reader access, with straightforward technique descriptions and examples of their practical use. In addition to the articles, there are one-page summaries for every technique, introductory summaries to groupings of related techniques, a complete glossary of acronyms, and a tabular comparison of the major features of all 50 techniques.

The 10 volumes in the Series on characterization of particular materials classes include volumes on silicon processing, metals and alloys, catalytic materials, integrated circuit packaging, etc. Characterization is approached from the materials user's point of view. Thus, in general, the format is based on properties, processing steps, materials classification, etc., rather than on a technique. The emphasis of all volumes is on surfaces, interfaces, and thin films, but the emphasis varies depending on the relative importance of these areas for the materials class concerned. Appendixes in each volume reproduce the relevant one-page summaries from the *Encyclopedia* and provide longer summaries for any techniques referred to that are not covered in the *Encyclopedia*.

The concept for the Series came from discussion with Marjan Bace of Manning Publications Company. A gap exists between the way materials characterization is often presented and the needs of a large segment of the audience—the materials user, process engineer, manager, or student. In our experience, when, at the end of talks or courses on analytical techniques, a question is asked on how a particular material (or processing) characterization problem can be addressed the answer often is that the speaker is "an expert on the technique, not the materials aspects, and does not have experience with that particular situation." This Series is an attempt to bridge this gap by approaching characterization problems from the side of the materials user rather than from that of the analytical techniques expert.

We would like to thank Marjan Bace for putting forward the original concept, Shaun Wilson of Charles Evans and Associates and Yale Strausser of Surface Science Laboratories for help in further defining the Series, and the Editors of all the individual volumes for their efforts to produce practical, materials user based volumes.

C. R. Brundle   C. A. Evans, Jr.

Preface to the Reissue of *Characterization of Compound Semiconductor Processing*

This volume was originally issued in 1995. At that time III–V based semiconductor devices had started to make a serious transition from the lab to the fab, being used in commercial modern communication and entertainment technology. This process has continued its course and II–V devices are now considered more "main stream" though, of course, the commercial volume of wafer processing remains small compared to silicon processing. The typical III–V materials problems and characterization issues in real world devices have not changed drastically since then, though, of course, new flavors have been added. Thus the original volume, covering materials and processing in GaAs, GaAlAs, InP and HgCdTe based devices, still provides insight into how materials characterization issues are dealt with. We are therefore initially reissuing it in its original form. This will then be followed up by individual updates and new chapters, which will be released as on-line downloads as they become available.

C. R. Brundle, C. E. Evans, and Gary E. McGuire, October 2009

Preface

This volume has been written to aid scientists and engineers working with compound semiconductor materials and devices in the selection and application of various analytical techniques. It highlights analytical problems that occur at all stages of materials or device processing (substrate preparation, epitaxial growth, dielectric film deposition, contact formation, and dopant introduction) and describes the application of a variety of analysis techniques in solving them. These techniques are illustrated in the investigation of surfaces, interfaces, thin films, defects, and impurities that affect material properties, processing and, ultimately, device performance. The techniques discussed are used as follow-up approaches to the simple electrical tests usually performed during device fabrication. The electrical tests are often insufficient, on their own, to pin down the origin of a problem, though they may indicate there is one.

This volume, and indeed the Materials Characterization Series, is intended to help the nonspecialist determine the best selection of techniques to analyze materials-related problems. Its purpose is to guide the nonspecialist by using examples of materials problems frequently encountered in compound semiconductor technology. The emphasis is placed on the materials problem rather than the details of the analysis technique, which is the basis of most other texts on analytical techniques. The volume is not intended to make one an expert in any of the individual materials characterization techniques. Further information to help solve materials-related problems may be obtained from the references at the close of each chapter.

Materials and processes used in the research, development and fabrication of GaAs, GaAlAs, InP and HgCdTe based devices provide examples of typical analytical problems. The application of a variety of characterization techniques gives the reader insight into how each individually, or in combination with other techniques, might be used to solve problems associated with these materials.

The chapters in this volume present aspects of the major materials areas of III–V and II–VI compound semiconductors, including the surface preparation and cleaning of substrate materials, epitaxial film growth by most of the major techniques, heterostructures, Schottky and ohmic contacts, dielectric films, and photon emitting and absorbing materials. The material, as well as the growth or deposition technique, is described. Many of the major analytical techniques are illustrated in each of the chapters, demonstrating the wide applicability of these tools.

This volume should be used in conjunction with the lead volume of the series, *Encyclopedia of Materials Characterization*, which defines boundary conditions for fifty widely used materials characterization techniques. Each technique description includes:

- a simple physical description of the technique

- the type of information to be obtained about a sample

- appropriate samples and required sample preparation

- limitations and hardware requirements with regard to spatial resolution, elemental specificity and sensitivity

- time required for an analysis

- sample degradation

- other important characteristics of the technique.

The fifty techniques discussed in the *Encyclopedia* are the most widely used for a broad range of materials problems. However, some of these techniques are seldom used in characterizing compound semiconductors, and some techniques specific to semiconductor characterization are not included. For these reasons, an appendix is provided in this volume that contains pertinent summary pages from the *Encyclopedia*, plus summaries of the important semiconductor-specific methods not covered in the *Encyclopedia*.

The Editors of this volume would like to thank Dick Brundle, the Series Editor, who helped beyond the call of duty in many ways. His patience and persistence have been invaluable in bringing this task to completion.

Yale Strausser   Gary E. McGuire

Contributors

Roger Brennan
Solecon Laboratories
Sunnyvale, CA

Spreading Resistance Analysis (SRA)

David Dickey
Solecon Laboratories
Sunnyvale, CA

Spreading Resistance Analysis (SRA)

Werner K. Götz
Xerox Palo Alto Research Center
Palo Alto, CA

Deep Level Transient Spectroscopy: A Case Study on GaAs

Noble M. Johnson
Xerox Palo Alto Research Center
Palo Alto, CA

Deep Level Transient Spectroscopy: A Case Study on GaAs

Walter Johnson
Prometrics Corporation
Santa Clara, CA

Sheet Resistance and the Four Point Probe

David C. Joy
The University of Tennessee-Knoxville
Knoxville, TN

Electron Beam Induced Current (EBIC) Microscopy

Thomas F. Kuech
The University of Wisconsin
Madison, WI

III–V Compound Semiconductor Films for Optical Applications

T. S. Low
Hewlett-Packard Laboratories
Santa Rosa, CA

Characterization of III–V Thin Films for Electronic Devices

George N. Maracas
Arizona State University
Tempe, AZ

Capacitance–Voltage (C–V) Measurements; Hall Effect Resistivity Measurements

J. N. Miller
Hewlett-Packard Laboratories
Palo Alto, CA

Characterization of III–V Thin Films for Electronic Devices

Philipp Niedermann
University of Geneva
Geneva

Ballistic Electron Emission Microscopy (BEEM)

Jon Orloff
University of Maryland
Washington, DC

Focused Ion Beams (FIBs)

David R. Rhiger
Santa Barbara Research Center
Goleta, CA

Other Compound Semiconductor Films

T. Sands
University of California-Berkeley
Berkeley, CA

S. A. Schwarz
Queens College
Flushing, NY

H. H. Wieder
University of California-San Diego
La Jolla, CA

Owen K. Wu
Hughes Research Laboratories
Malibu, CA

Chuck Yarling
Prometrics Corporation
Santa Clara, CA

Contacts

Contacts

Dielectric Insulating Layers

Other Compound Semiconductor Films

Sheet Resistance and the Four Point Probe

Characterization of III–V Thin Films for Electronic Devices

J. N. MILLER and T. S. LOW

Contents

1.1 Introduction

The process of producing III–V semiconductor electronic devices has moved out of the laboratory and into commercial markets that impact many of our advanced modern communication and entertainment technologies, including cellular telephones, pocket pagers, and direct satellite broadcast receivers. Both individual transistors and integrated circuits (ICs) are fabricated in mainly GaAs-based III–V semiconductors. Passive elements such as resistors, capacitors, and even inductors are needed to make III–VICs, but the transistor gain element is the engine for these ICs. The three predominant transistor device structures for such devices are the metal-semiconductor field effect transistor (MESFET), the modulation-doped field effect transistor (MODFET), and the heterojunction bipolar transistor (HBT). At present, the MESFET accounts for the largest fraction of devices produced. In this chapter, we demonstrate how surface analytical techniques can and have been used to solve real world problems in producing III–V electronics.

Such electronic devices are fabricated in small regions of a III–V semiconducting film that have been made electrically conducting, either by implantation or in-diffusion of dopants, or else by epitaxial growth of doped semiconductor films. In contrast to silicon technology that makes wide use of p–n junctions and trench etching to isolate electronic devices from each other, the III–V technologies usually isolate electronic devices by using nonconducting semiconductor material. Several of the larger bandgap III–V semiconductors can be rendered nonconducting enough for isolation purposes (i.e., semi-insulating) by controlling residual deep and shallow level concentrations through crystal growth conditions, by intentional doping with deep centers, and by ion implantation. The isolated electronic devices are then connected to each other with metal transmission lines, which are in turn connected to conducting semiconductor regions with either low-resistance ohmic contacts or stable, high-quality Schottky contacts. In this chapter, we focus on the surface and thin film characterization of both bulk substrate and epitaxially grown material, and the formation of both conducting and semi-insulating regions using ion implantation.

1.2 Surface Characterization of GaAs Wafers

Typically, the beginning of the process of making devices involves removing wafers from their packaging materials. Although the quality of wafer packaging has improved over the years, the surfaces of as-received wafers are not atomically clean. The wafers have native oxides (1 to 10 nm thick), hydrocarbons adsorbed from handling and from the plastic shipping bags, and assorted other chemical contaminants. The presence of such contaminant films can be detected, and their thickness measured, using standard single-wavelength ellipsometers.[1] The elemental composition of these films is sometimes measured by Auger electron spectroscopy (AES) and X-ray photoelectron spectroscopy (XPS) (which are discussed at greater length below). A certain amount of chemical information about the contaminant films can be deduced from which wet chemical treatments do or do not remove them, as measured via ellipsometry.

The surfaces of the as-received wafers are also contaminated with particles (from a few to a few hundred per wafer between 0.5 and 10 µm in size). There are several commercially available instruments that count particles on blank (and even patterned) wafers and categorize them by size. Most of these instruments work by rastering a focused laser beam over the wafer. In the absence of particles, the laser beam is simply specularly reflected off the wafer in a well-defined direction. A detector is positioned so as to not see this specularly reflected beam. When the beam strikes a particle, light is scattered in all directions, and the detector senses a flash of light. The intensity of the flash is calibrated with known-size particles, and the machine can parse the flashes detected on the wafer into a histogram displaying the number of particles versus their size. These instruments are also useful for measuring the number and size of morphology defects in III–V semiconductor epitaxial layers, and of particles added during fabrication processes. These defects and particle contaminants must be controlled to achieve acceptable III–V IC yields; thus, particle counters provide useful feedback for the qualification of substrate and epitaxial material, and for controlling individual processes during wafer fabrication.[2]

The first aspect of surface analysis critical to the production of devices is the measurement of the surface finish of the starting wafers. The most widespread technique is the optical examination of the surface using a Nomarski interference contrast optical microscope. Nomarski microscopy is indispensable in evaluating surface defects and particles, scratches, polishing texture, and saw damage. It has the disadvantage of all image-based techniques in that, although the information density is high, it is difficult to make any quantitative measurements or comparisons of surface morphology.

Even in substrates with completely featureless, smooth morphology there can be mechanical damage beneath the surface (e.g., from sawing, polishing, or improper handling) that extends as much as several micrometers below the surface. The damage can include a variety of point and extended defects, as well as microscopic cracks. Some or all of this damage may be removed by the wet etches (removing one or more micrometer of the substrate surface) frequently used prior to epitaxial growth. However, many ion implantation processes do not involve such etches, so they are more sensitive to such damage. Etching techniques have been developed[3] that purport to decorate this subsurface damage in GaAs wafers so that it is visible in the surface morphology. A host of characterization techniques have been applied to measuring subsurface damage—including laser light scattering (analogous to the particle counters described above), photoreflectance, spectroscopic ellipsometry (i.e., ellipsometry versus wavelength), cathodoluminescence (CL), and electron beam induced current—but no one technique has stood out as superior to the others. Fortunately, the quality of commercially available GaAs and InP substrate surface polish has improved to the point that characterization of subsurface damage is of less interest than it was ten years ago.

Dislocations

Dislocations are extended crystalline defects, and the dislocation density is a figure of merit of the substrate material. Dislocations, and their motion through the crystal during bulk growth, contribute electrically active deep level defects that can influence the performance of nearby III–V electronic devices. Dislocations also give rise to macroscopic defects in surface morphology, such as growth defects in epitaxial films and features that form via nonuniform etching around the dislocation during processing. The dislocation density can also have an indirect impact on the local activation of dopants in ion-implanted material by its influence on local native defects (e.g., EL2 in GaAs) and/or residual donor or acceptor concentrations. The threshold voltages of ion-implanted MESFETs have been reported[4] to correlate with the density and proximity of threading dislocations in the substrate. The substrate dislocation density is often measured by counting the density of surface "etch pits" after the wafer has been subjected to a delineation etch. There are a large number of etches which delineate dislocations, including molten KOH. As well as chemical techniques, dislocations have been studied by electron microscope techniques.[5–7]

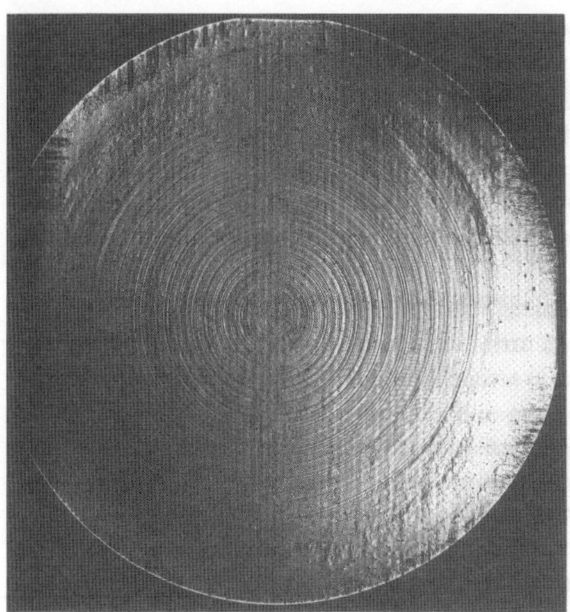

Figure 1.1   An XRT image of a typical GaAs wafer grown by the LEC technique. The dark spots and lines correspond to crystalline defects (mostly dislocations). The ring-like structure is caused by convective thermal nonuniformities in the molten GaAs, which results in small changes in the lattice constant due to small shifts in stoichiometry and/or impurity segregation.

A nondestructive technique for measuring surface dislocation densities is X-ray topography (XRT). In substrates with defect densities less than a few hundred per square centimeter, XRT can produce a clear image of the dislocations. The XRT contrast is achieved when the diffracted intensity in a particular X-ray reflection is changed as a result of changes in the local lattice parameter, which can be caused by a dislocation or compositional striation or some other condition that perturbs the local bond lengths in the wafer. Figure 1.1 is an X-ray topograph of a liquid encapsulated Czochralski (LEC) GaAs substrate.

Surface Composition and Chemical State

The composition and chemical state of the surface is of great importance for subsequent processing. For example, proper nucleation in epitaxial growth must begin with an atomically clean surface. As mentioned previously, the wafers are exposed to room air and to hydrocarbon contamination from the packing material and other contaminants. After unpacking, and prior to epitaxial growth or other processing, the wafers are typically "cleaned." They are "degreased" to remove hydrocarbon residue; this is followed by acetone and isopropanol rinses. The wafer may then be etched, in a hydrogen peroxide and water solution, with either ammonium hydroxide or sulfuric acid. The purpose of this etch is to remove the potentially damaged near surface of the crystal, and to leave the wafer surface in a uniform state of oxidation.[8] Fortunately, in recent years substrate manufacturers have improved their polishing and packaging techniques so as to minimize the need for any surface treatment prior to some processing steps.

Electron spectroscopies are likely to yield the most information about the composition and chemical state of the wafer surface. The two most common techniques are AES[9] and XPS.[10] The extreme surface sensitivity of these techniques makes them well suited for measuring the surface chemical state. AES and XPS are particularly sensitive to the surface concentration of carbon and oxygen, two of the most common surface contaminants. These techniques measure the elemental identity and surface concentration of the surface contaminant within a few to a few tens of angstroms of the surface via the atom's unique core transition energies. Some of these transition energies are slightly shifted, depending on the atom's charge and its local chemical environment. In some cases, it is possible to use these energy shifts to determine the chemical compounds present on the surface (as opposed to simply getting atomic surface concentrations). Applications of the AES and XPS techniques are discussed in many places in this book series.

In Rutherford backscattering spectrometry (RBS),[11] the recoil energy of He^{2+} ions is measured and used to deduce the mass and location of atoms in the lattice. Since ion-scattering cross sections are absolute functions of energy, the concentration of species can be measured without calibration standards. High dosage implants of As or other heavy species are measured with the greatest sensitivity and are used to calibrate the ion implanter dosage measurement equipment. RBS is not generally useful for measuring the low implantation levels typical in III–V device fabrication. RBS is particularly effective in measuring metallic ohmic-contact depth profiles. Ohmic-contact metallizations are often specifically chosen to be highly reactive or have low eutectic melting temperatures. Owing to this, the ion beams typically used in AES or secondary ion mass spectrometry (SIMS) profiling cause ion beam mixing and preferential sputtering; thus the resulting scrambled profile may not reflect the original interfacial metallurgy. RBS does not suffer from this limitation, but it provides no chemical state information, unlike AES or XPS.

RBS can also be used to measure the crystalline quality in the near-surface region by measuring the channeling of ions into the interstices of the lattice. In the channeling mode, the scattered ion intensity is measured as a function of tilt angle. The intensity reaches a local minima each time the crystallographic axis of the material is aligned with the ion beam. The degree of channeling in a sample is a measure of the crystalline quality of the near-surface region.

SIMS in the "static" mode[12] has also been used to study the surface. Static SIMS relies on extremely low average primary ion-beam currents to effectively sample the chemistry on the surface and then mass-analyze it. The average beam current densities are chosen to be low enough that a primary ion is unlikely to strike a region already bombarded by a previously arriving primary ion. By measuring the charge-to-mass ratio of the surface species, a model of the surface prior to sputtering can be constructed.

SIMS in the "dynamic" mode—where the atomic concentrations are determined as a function of sputtered depth by using higher primary ion-beam currents—is used primarily for depth profiling through thin films, rather than examining the outermost surface. Its information content is thus similar to RBS, but its depth resolution and absolute sensitivity level are better. It does suffer from ion-beam induced mixing, as mentioned above. SIMS is discussed again in the next section. Full descriptions of all these techniques, and their applications, may be found in the lead volume of this series, *Encyclopedia of Materials Characterization*.

1.3 Ion Implantation

The fabrication of FETs and other III–V semiconductor electronic devices requires the formation of an *n*- or *p*-type conducting channel. The conducting channel is often formed by implanting high energy ionized *n*- or *p*-type dopant atoms (e.g., Si or Be, respectively) into the substrate.[3] Predicting and characterizing how device characteristics are influenced by ion-implantation parameters and by post-implantation anneal conditions is a subspecialty of its own, and only a few simple examples can be given here. The reader can find more details on this broad subject in References 3 and 4 and references therein.

The performance of all semiconductor devices depends upon controlling the precise location and concentration of electrically active dopant atoms and of native and extrinsic deep centers that may be present. Implantation-based III–V electronic devices are no exception to this rule. By adjusting the implantation parameters (e.g., ion species, ion dose, ion energy, ion incidence angle relative to crystal axes), the profile of dopant atomic concentration versus depth can be controlled. Typical implantation energies range from the tens of kiloelectronvolts to 1 MeV, and these energies correspond to ion penetration depths of about 10 to 10^4 nm in GaAs. Because Ga and As atoms are much more massive than Si, substantially higher implant energies are required for GaAs than for Si to achieve a given penetration depth. The ions come to rest in the semiconductor lattice with a depth profile that can be predicted by well-established models.[13]

The implanted dopant atoms are generally not on electrically active lattice sites, and there is a large concentration of defects in the semiconductor after the implant. These defects were formed by the ions as they crashed through the crystal lattice and came to rest. The defects include misplaced column III or column V atoms (e.g., Ga or As interstitials or vacancies), and complexes of such defects with each other and with the implanted dopant atoms. For high-dose implants, the concentration of such defects can be so high that the crystal is essentially amorphous. Many of these defects are electrically active deep levels and can thereby compensate the dopant atoms that are present.

Fortunately, most of the implanted dopant ions can be moved to electrically active substitutional sites, and most of the implantation-induced point defects can be removed by a post-implant anneal. Several parameters are optimized and carefully controlled for such anneals (e.g., anneal temperature versus time, anneal caps, arsenic or other column V element overpressures, etc.). Besides rendering the dopant atoms electrically active and removing implant-induced defects, the anneal can cause the dopant atoms to diffuse. This diffusion can substantially change the dopant atom concentration profile from its as-implanted shape.

The profile of the dopant atom concentration versus depth is often measured by SIMS, as mentioned previously.[14] The SIMS measurement can be used before and after the post-implant anneal to measure the as-implanted profile and the change in the profile induced by the post-implant anneal. It is important to realize that SIMS measures only the atomic concentration of the dopant atoms, not the electrically active concentration, which can be quite different. What SIMS actually measures is the count rate for a specific charge-to-mass species (interpreted as a specific ionic species) as a function of time while the surface of the sample (including dopant atoms) is steadily sputtered away. The measured charge-to-mass count rates versus sputter time can be converted to atom concentration versus depth by calibrating count rates on standard samples,[15] and by measuring the depth of the sputtered crater in the sample by profilometer. The applications of SIMS are described in more detail elsewhere in this series.[16] An example of a SIMS profile of Zn concentration versus depth is shown in Figure 1.2. The sensitivity of SIMS to many ionic species is in the parts-per-billion range, which makes SIMS useful for measuring dopant profiles over nearly the entire range of doping used in III–V semiconductors. Accuracies of dopant concentrations of ±5% can be obtained in some SIMS applications, but typical accuracies are about ±50%.

Another figure of merit of the ion-implanted layer after annealing is the so-called implant activation. This is simply the ratio of the electrically active dopant atoms (i.e., the number of free electrons or holes per square centimeter) to the total implantation dose (e.g., the number of implanted Si or Be atoms per square centimeter, as measured from the Faraday cup in the ion implanter). The compensating implantation-induced defects are generally not completely removed by the anneal, and there may be some self-compensation by the dopant (e.g., for Si, which can form both donors and acceptors, depending on the lattice site it occupies); so the implant activation is generally less than unity. The implant parameters, the anneal conditions, and the bulk substrate material are all optimized to make the activation close to unity and to make it as uniform as possible across the wafer.

The usual techniques to measure implant activation include contactless conductivity probes[†] (nondestructive and accurate to about 1%). Capacitance–voltage

---

† From, for example, Tencor Co., Mountain View, CA.

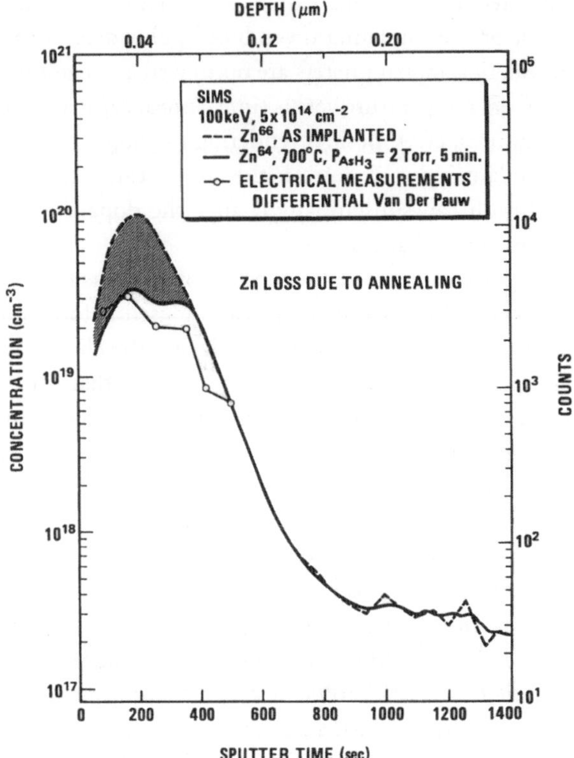

Figure 1.2   A SIMS depth profile for zinc-implanted GaAs. Two isotopes were used to accurately quantify the loss of zinc-64 isotope due to annealing. The zinc-66 isotope was implanted at the same dose after annealing. The gray region indicates the lost zinc. Also shown is a differential electrical measurement in good agreement with the SIMS measurement.

measurement (C–V) via mercury probe or evaporated Schottky contacts, and the related technique of electrochemical C–V are also sometimes used to measure implant activation. They measure the electrically active doping profile (as distinguished from the atomic concentrations measured by SIMS), but these C–V techniques have several significant disadvantages. They contact and destroy or contaminate the measured area on the wafer, and they are not very accurate (measured concentration is very sensitive to Schottky contact area and to Schottky barrier quality). They also suffer gross inaccuracies when measuring doping profiles near the boundary of conducting and insulating layers (e.g., the implant profile near the bottom of an implanted FET channel). Unfortunately, this part of the profile is often of great interest for predicting device performance.

The electrically active dopant profile can also be measured by differential Hall effect. This is done by repeatedly etching small amounts of the implanted layer

                CHARACTERIZATION OF III–V THIN FILMS . . .   Chapter 1

away, and then making a Hall effect measurement. The doping and mobility profile versus depth can be deduced from such a measurement sequence. Differential Hall effect is a destructive and very time consuming technique, but it can give an accurate profile near the bottom of an FET channel. The most accurate and satisfactory way of measuring the electrically active dopant profile in an implanted FET layer is by measuring the $I_{ds}(V_{ds}, V_{gs})$ characteristics of a long gate FET processed on the implanted material. This obviously has the disadvantage of slow feedback about the implant, but it is widely used because it reliably predicts actual device performance.

Transmission electron microscopy (TEM) is sometimes used to measure the density and distribution of extended defects that remain in the implanted layer after the post-implant anneal.[17] The anneal causes these defects to aggregate into dislocations, stacking faults, and other extended crystalline defects which are visible in a TEM image. However, the TEM image provides no information about the point defects (vacancies, dopant-antisite defect complexes, etc.). Many of the extended defects and point defects are electrically active, and contribute to incomplete dopant activation. The TEM technique is not used extensively because it is destructive, because it samples an extremely small (and potentially nonrepresentative) region of the film, and because sample preparation is difficult and tedious and can often introduce crystalline defects of its own that confuse the interpretation of the TEM images.

The deep centers associated with the implantation damage (or similarly, those associated with epitaxial growth) often introduce unwanted transients in the device response, which are often quantified using deep level transient spectroscopy (DLTS). DLTS techniques can either be applied directly to the implanted or epitaxial layer (destructive), or they can be performed on processed devices (nondestructive). Often the most useful data are obtained from DLTS studies of actual devices, but the DLTS data obtained from simple test devices fabricated directly on the implanted or epitaxial layer have the virtue of usually being easier to interpret. The application of DLTS techniques is another subspecialty that goes beyond the scope of this chapter, and the reader is referred to Chapter 6 of this volume and Reference 18. The two versions of DLTS in most common use for characterizing III–V electronic device films (including both implanted layers and epitaxial films) are capacitance DLTS (CDLTS) and conductance DLTS (GDLTS). CDLTS measures (via depletion capacitance transients) the concentration and emission or capture time constants associated with deep centers in a depletion region, such as that formed by a p–n junction in a diode or heterojunction bipolar transistor or that formed under a Schottky contact such as the gate of a long gate FET. In contrast, GDLTS measures (e.g., via FET channel conductance transients or HBT collector current transients) the concentration and emission/capture time constants of deep centers in the active region of the device under test. These two techniques are also sometimes used to measure process-induced deep levels and to characterize transients associated with the filling and emptying of surface states (e.g., on the semiconductor surface between the gate and the source or drain ohmic contacts).

Ion implantation is also important for III–V electronic devices, particularly for the production of ICs, because it can be used for electrical isolation of devices. The III–V semiconductors, especially those with relatively larger bandgaps than Si (e.g., GaAs, InP, and AlGaA) can be made nonconducting ("semi-insulating") by introducing a sufficient number of deep level donors and/or acceptors to compensate the shallow donors or acceptors that are present, thereby pinning the Fermi level near the mid-bandgap. This produces material that is suitably nonconductive for device isolation. Such semi-insulating semiconductor substrate material is routinely produced by bulk crystal growth. The bulk material is made semi-insulating by controlling residual deep and shallow donor and acceptor concentrations and/or by introducing deep centers via doping with transition metals like chromium or iron. Sometimes epitaxial films are also grown to be semi-insulating by doping with deep centers (such as O in AlGaAs or Fe in InP). Formerly conducting ion implanted or epitaxially grown films can also be rendered semi-insulating for isolation by ion implantation (of protons, boron, oxygen, etc.), and the mechanism is the same.[3] An older alternative technique is mesa isolation, where the conducting layer or layers are simply etched away in the areas to be isolated, leaving only the semi-insulating substrate. Mesa isolation has the disadvantage that it is nonplanar, so step coverage can be problematic for IC yields, and the electrical isolation is poorer than for implant isolation for small device spacings.[19] Unfortunately, mesa isolation may be the only possibility for the emerging InGaAs channel devices, because the narrow bandgap of InGaAs will make implantation isolation less effective than for GaAs.

The structural defect and atomic concentration profile characterization techniques for the isolation implants are the same as for conducting channel implants, but the figures of merit for an isolation implant layer and the corresponding electrical characterization techniques are quite different.[19] A low value for the sheet conductivity of a blank wafer may be measured after the isolation implant and after whatever post-implantation anneal may have been done. However, this is a fairly crude test of the quality of the isolated layer, and low sheet conductivities may be measured for a wide range of implantation and anneal conditions, many of which will correspond to unacceptably poor isolation in a realistic IC environment. One of the reasons for this is that the semi-insulating material has a dramatically nonlinear current-voltage characteristic, especially at high electric fields.[20] The voltage at which large currents start to flow between nominally isolated conducting regions on an IC increases with the concentration of empty deep level traps in the isolated material, and decreases steeply with the spacing between the conducting regions. The quality of the isolation implant can be tested by fabricating test devices on the isolated material (e.g., transmission line patterns of typical IC intercontact spacings), but the isolation quality is somewhat surface sensitive, and so it depends on the details of the IC process. The quality of the isolation implant, therefore, is most realistically tested via test structures of typical IC design rule spacings on the actual processed wafer (e.g., ohmic to ohmic, and ohmic to Schottky contact structures with variable spacings a few to a few tens of micrometers).

1.4 Epitaxial Crystal Growth

The conducting regions of semiconductor material required to fabricate III–V electronic devices are often produced by epitaxial crystal growth. Molecular beam epitaxy[21] (MBE) and organometallic vapor phase epitaxy[22] (OMVPE) have emerged as the primary growth techniques for III–V electronic device fabrication. Crystal growth by MBE is accomplished by the evaporation of column III atoms (e.g., Ga or Al), column V atom(s) (e.g., As as As_4), and n or p dopant atom(s) (e.g., Si or Be, respectively) from source furnaces containing the elemental form of each. When the "molecular beams" from these sources are directed toward an atomically clean single crystal substrate at a controlled elevated temperature, they form an "epitaxial" layer (i.e., a single crystal layer with the same atomic spacing and orientation as those of the underlying substrate). Epitaxial crystal growth by OMVPE uses organometallic compounds (e.g., trimethylgallium [TMGa] or triethylaluminum [TEAl]) to transport column III elements, and uses hydrides (e.g., AsH_3 or PH_3) to transport the column V elements. The substrate is held at a controlled elevated temperature, and these compounds pyrolytically decompose at or just above the substrate surface, and react on the substrate to form the III–V epitaxial layer. Dopants (e.g., Si as SiH_4 or C as CCl_4) are also transported to the substrate and incorporate into the epilayer from the gas phase.

Reflected high energy electron diffraction[23] (RHEED) has been used extensively since Cho's early work[24] to characterize the MBE growth surface. An electron beam at or near 10 keV strikes the crystal surface at a grazing angle. The electrons are diffracted by the surface lattice and form beams which result in streaks or lines on a phosphor screen. The spacing between these lines is proportional to the surface reciprocal lattice vector.

RHEED images of the surface have proven to be a good way to verify the cleanliness and crystallinity of the starting surface. Wafers are introduced into the MBE system with a controlled oxide typically one or two monolayers thick. The oxide is desorbed by heating the GaAs wafer in an As beam to approximately 600 °C. The oxide is amorphous and results in only a diffuse RHEED pattern. At the time of the oxide desorption, the diffuse pattern gives way to a well-defined 2 × 4 pattern, indicating the surface crystallinity and that the surface is atomically clean, at least in some areas. The stoichiometry of the GaAs surface is also a strong function of the ratio of arriving Ga and As. The RHEED pattern shifts from 2 × 4 to 4 × 2 as the surface shifts from an As-stabilized surface to an As-depleted surface. Owing to this, the RHEED patterns undergo a phase transition when the surface V:III ratio drops to less than one.

The intensity of the RHEED streaks oscillates[25] as a function of time upon the initiation of growth, under the proper conditions. These RHEED oscillations are periodic with the growth of monolayers of material. As a consequence, measuring RHEED oscillations is a convenient and accurate method with which to calibrate growth rates.

In recent work by Pashley et al.[26] the GaAs surface has been imaged directly using scanning tunneling microscopy (STM). They prepared the GaAs to give a 2 × 4 As-stable surface, and interpreted their surface images to be indicative of three pairs of As dimers and one missing dimer as the atom configuration filling the 2 × 4 surface unit cell. Though the STM technique requires equipment not widely available, the power of this technique to image surfaces directly will likely increase its use and spawn related techniques.

AES, which is discussed earlier, has been extensively used to characterize the MBE growth surface. Measurements of surface cleanliness and the surface As-to-Ga ratios have been made by AES and XPS. Arthur[27] made extensive use of Auger in his early work on MBE growth.

OMVPE crystal growth is performed with gases flowing in the viscous flow regime. As a consequence of the high pressure, electron probes of the surface, such as RHEED, XPS, and AES, are not usable. Instead, optical techniques have been developed to monitor the growth surface.

Ellipsometry and other measurements of the intensity and phase of light reflected from a wafer surface have been shown to be useful in measuring growth rate and Al composition in AlGaAs alloys. Spring Thorpe and Majeed[28] developed a temperature oscillation technique using the Fabry–Perot resonator that exists between a growing GaAs surface and an underlying AlAs/GaAs heterointerface. The refractive index discontinuities at these two interfaces produce oscillations in the apparent temperature, as measured by a pyrometer, as the GaAs grows. The period of these oscillations can be translated into thickness and growth rate information. Recently, Robbins et al.[29] reported using a similar technique in an OMVPE reactor. Workers at CNRS[30] in the early 1980s reported the design of an OMVPE reactor with an ellipsometer designed into the reactor which could simultaneously measure thickness and composition of AlGaAs layers.

Colas[31] has integrated reflectance difference spectroscopy into an OMVPE reactor. The reflectance difference (RD) technique measures the difference between the reflectance in the [1,1] and the orthogonal [1,−1] surface direction. The ratio of these two signals oscillates with the same phase relationship as the growth of a GaAs monolayer. So, RD can be used in the same way RHEED is used in MBE to calculate and calibrate growth rates.

In situ X-ray measurements have also been made in an OMVPE reactor specifically designed for grazing incidence X-ray measurements.[32] X-ray scattering results indicate that the surface during OMVPE growth may have a 2 × 4 surface periodicity.

After the wafer is removed from the growth apparatus, a number of characterization techniques can be employed to verify that the structure and doping in the epilayers correspond to the desired values. Electrical measurements of sheet resistance, Hall-effect measurements of electron mobility and carrier concentration, C–V and DLTS measurements of the carrier concentration and deep level concentration are used to characterize the electrical behavior of the fabricated structures.

X-ray measurement techniques are also widely used to measure crystallinity and lattice parameters. X-ray measurements map the crystal's reciprocal lattice, or spatial Fourier representation of the lattice. In AlGaAs, X-ray diffraction measurements are taken for the Al mole fraction. In the growth of other alloy systems (e.g., InGaAs), X-ray measurements can be used to identify the correct composition for lattice matched growth, pseudomorphic layers, and the degree of tetragonal distortion induced by certain growth conditions. An excellent review of these X-ray techniques has been written by Picraux et al.[33]

Epitaxial crystal growth also suffer from particles and defects generated during growth. The characteristic "oval defects" of MBE can be very numerous (e.g., >1000/cm^2), and these defects have a detrimental effect on device, and especially IC, yield. Modern Ga cells have reduced these defect densities to as low as 10/cm^2, but they remain a major obstacle to III–VIC fabrication by MBE epitaxial material. Oval defects are believed to be caused by an agglomeration of surface Ga atoms, which results in a liquid-solid growth interface. The defect is bounded by {111} stacking faults in most cases.[34] Oval defects can nucleate on particles present on the wafer surface prior to growth. Shinohara has demonstrated changes in FET threshold voltage caused by such defects under the gate.[34]

1.5 Summary

In this chapter, we have presented a number of key examples of how surface analytical techniques are used to solve real world problems in producing III–V electronic devices and ICs. We have also tied these examples to some of the basic processing steps involved in producing these types of materials and devices.

References

1 D. E. Aspnes. In *Optical Properties of Solids: New Developments*. North-Holland, Amsterdam, 1976.

2 B. J. Tullis. *Microcontamination*. 67, Nov. 1985.

3 P. J. Caldwell, W. D. Laidig, Y. F. Lin, C. K. Peng, T. J. Magee, and D. Leung. *J. Appl. Phys.* **57**, 984, 1985; *GaAs FET Principles and Technology* (J. V. DiLorenzo and D. D. Khandelwal, Eds.) Artech House, Dedham, MA, 1982.

4 H. Matsura, H. Nakamura, Y. Sano, T. Egawa, T. Ishida, and K. Kaminishi. *GaAs IC Symposium Technical Digest*. IEEE, New York, 1985, p. 67; *Gallium Arsenide Technology*. (D. K. Ferry, Ed.) H. W. Sams & Co., Indianapolis, 1985.

5 K. H. Kuesters, D. C. DeCooman, and C. B. Carter. In *Thirteenth International Conference on Defects in Semiconductors*. (L. C. Kimerling and J. M. Parsey, Eds.) AIME, New York, 1985.

6 P. Hirsh, A. Howie, R. B. Nicholson, D. W. Pashley, and M. J. Whalen. *Electron Microscopy of Thin Crystals*. R. E. Keiger, Malabar, FL, 1977.

7   H. J. Leamy. *J. Appl. Phys.* **53**, R51, 1982.

8   A. J. SpringThorpe and P. Mandeville. *J. Vac. Sci. Tech.* **B4**, 853, 1986.

9   P. H. Holloway. *Surf. Sci.* **54**, 506, 1976.

10 C. R. Brundle and A. D. Baker. *Electron Spectroscopy: Theory, Techniques and Applications.* Vols. 1–4, Academic Press, London, 1981.

11 W. K. Chu, J. W. Mayer, and M.-A. Nicolet. *Backscattering Spectrometry.* Academic Press, London, 1978.

12 A. Benninghoven. *Surf. Sci.* **53**, 596, 1975.

13 J. F. Gibbons, W. S. Johnson, and S. W. Mylroie. *Projected Range Statistics,* 2nd ed. Dowden, Hutchinson and Ross, Stroudsburg, PA, 1975.

14 J. B. Clegg, A. E. Morgan, H. A. M. deGrefte, F. Simondet, A. Huber, G. Blackmore, M. G. Dowsett, D. E. Sykes, C. W. Magee, and V. R. Deline. *Surf. and Interface Anal.* **6**, 162, 1984.

15 J. N. Miller. In *Proceedings of the Sixth International Conference on Ion Beam Analysis.* (W. A. Lanford, I. S. T. Tsong, and P. Williams, Eds.) North-Holland, Amsterdam, 1983.

16 P. K. Chu. "Dynamic SIMS." In *Encyclopedia of Materials Characterization.* (C. R. Brundle, C. A. Evans, Jr., and S. Wilson, Eds.) Butterworth-Heinemann/Manning, Boston, 1992.

17 *Semi-Insulating III–V Materials.* (S. Makram-Ebeid and B. Tuck, Eds.) Shiva, Cheshire, 1982.

18 *Point Defects in Semioconductors: I. Theoretical Aspects, and II. Experimental Aspects.* Series in Solid State Sciences, Springer-Verlag, New York, 1983.

19 D. C. D'Avanzo. "Proton Isolation for GaAs Intergrated Circuits." *IEEE Transactions on Microwave Theory and Techniques.* **MTT-30** (7), 955, July 1983.

20 M. A. Lampert, and P. Mark. *Current Injection in Solids.* Academic Press, New York, 1970.

21 *The Technology and Physics of Molecular Beam Epitaxy.* (E. H. C. Parker, Ed.) Plenum, New York, 1985.

22 G. B. Stringfellow. *Organometallic Vapor-Phase Epitaxy.* Academic Press, San Diego, 1989.

23 P. R. Pukite, J. M. Van Hove, and P. I. Cohen. *J. Vac. Sci. Tech.* **B2**, 243, 1984.

24 A. Y. Cho. *J. Vac. Sci. Tech.* **8**, 531, 1971.

25 J. H. Neave, B. A. Joyce, P. J. Dobson, and N. Norton. *Appl. Phys.* **A31**, 1, 1983.

26 M. D. Pashley, K. W. Haberern, and W. Friday. *Phys. Rev. Lett.* **60**, 2176, 1988.

27 J. R. Arthur, Jr. *J. Vac. Sci. Tech.* **10**, 136, 1973.

28 A. J. Spring Thorpe and A. Majeed. *J. Vac. Sci. Tech.* **B8**, 266, 1990.

29 V. M. Robbins, E. C. Jones, J. N. Miller, D. E. Mars, and A. S. Wakita. Technical Program from 5th OMVPE Workshop. TMS, Warrendale, PA, 1991.

30 G. Laurence, F. Hottier, and J. Hallais. *J. Cryst. Growth.* **55**, 198, 1981.

31 E. Colas. *Proceedings.* International Conference on Modulation Spectroscopy (F. H. Pollak, M. Cardona, and D. E. Aspnes, Eds.), SPIE, Bellingham, WA,` 1990, p. 94.

32 D. W. Kisker, G. B. Stephenson, P. H. Fuoss, F. J Lamelas, S. Brennan, and P. Imperatori. *J. Crystal Growth.* **124**, 1, 1992.

33 S. T. Picraux, B. L. Doyle, and J. Y. Tsao. *Strained Layer Superlattices.* (Willardson & Beer) Semiconductors and Semimetals Volume, 33rd ed., Academic Press, San Diego, 1991, p. 170.

34 M. Shinohara, T. Ito, K. Wada, and Y. Imamura. *Jap. J. Appl. Phys.* **23**, L371, 1984.

III–V Compound Semiconductor Films for Optical Applications

THOMAS F. KUECH

Contents

2.1 Introduction

Optical devices have become an important part of our technology. The light emitting diode (LED) is an integral component of many displays. The solid state laser is used in communication systems, connecting our phones and data processing equipment through fiber-optic cables. Consumer products, such as the compact disk player, have benefited from the availability of inexpensive, compact, and reliable solid state light emitters and detectors. Other optical components required in an optical system are now available in solid state form. Detectors, modulators, and waveguides can be formed through the application of modern epitaxial growth and semiconductor-processing techniques. Compound semiconductors, such as GaAs, $Al_xGa_{1-x}As$, InP, and $In_xGa_{1-x}As_yP_{1-y}$, are all direct bandgap materials suitable for such optical device structures. The availability of multicomponent semiconductor alloy systems provides independent control of both bandgap and lattice parameter. The bandgap controls the wavelength regime over which the material interacts

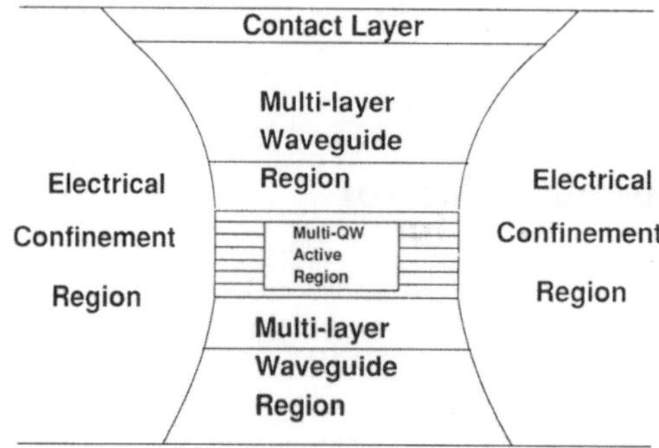

Figure 2.1   A double heterostructure laser is typical of many of the advanced multilayer optical devices being developed and manufactured. This structu re contains separate layers for the optical waveguiding and electrical current confinement required for efficient lasing. Multiple epitaxial steps are often employed to regrow specialized regions onto an existing etched and patterned structure.

strongly with light. The choice of lattice parameter is important in the growth of high-quality material on readily available substrates.

A distinguishing feature of compound semiconductor devices, in general, over Si-based technology lies in the formation of the active device region. In Si-based devices, the active region is formed through the modification of the near-surface region of a single-crystal Si wafer. Advanced compound semiconductor devices are formed, however, through the process of epitaxial crystal growth. Thin regions of semiconductors are deposited onto an existing substrate, typically GaAs or InP. Through controlled deposition onto this substrate, multiple semiconductor layers differing in composition and doping are deposited on the substrate in sequence such that the crystallographic orientation is maintained. This process of epitaxial growth is central to forming the complex multilayered structures needed in laser fabrication. A laser structure, shown schematically in Figure 2.1, contains many individual layers and can possibly entail multiple growth steps. All the individual layers and their interfaces must work properly to produce a high-efficiency laser. The deposited film must have a lattice parameter close to that of the substrate to prevent disregistry of the atoms as the film grows. Such disregistry eventually leads to the formation of strain-relieving defects, such as dislocations, which can strongly degrade the device performance. Pseudomorphic growth, shown schematically in Figure 2.2, allows for the strained epitaxial growth of thin lattice-mismatched layers. This growth continues up to a "critical" thickness beyond which defects are formed to relieve the built-in strain.[1] In either lattice-matched or -mismatched growth, materials properties are largely determined by alloy composition and growth conditions. These materials properties are dependent on the structural characteristics of the film, such as thickness and morphology, which must be monitored and controlled.

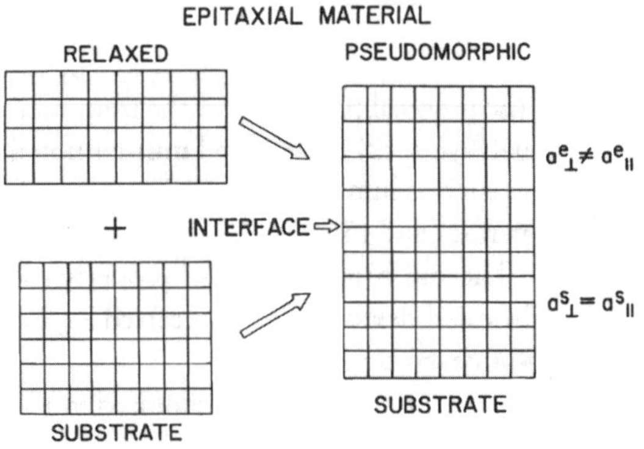

Figure 2.2 The growth of pseudomorphic layers expands the range of possible materials useful in optical device structures. The lattice mismatch is accommodated by increasing the strain in the overgrown layer. This type of growth will continue up to a "critical" thickness after which strain-relieving defects such as dislocations are formed.

Several forms of crystal growth are used in the formation of these multilayer structures: liquid phase epitaxy (LPE), vapor phase epitaxy (VPE), metal-organic vapor phase epitaxy (MOVPE), and molecular beam epitaxy (MBE). Most LED structures have been formed using LPE and VPE. New material structures, such as quantum well lasers, are being primarily grown by MOVPE and MBE. All these growth techniques, generally, do not directly monitor the film formation during the actual growth process. As a result, the formation of a multilayered structure involves an iterative process of growth, measurement, and then growth based on the refined parameters. This process places a heavy burden on the characterization of single-layer structures as well as multilayer structures of increasing complexity.

Most optical devices utilize the standard semiconductor device fabrication techniques—ion implantation, diffusion, and dielectric deposition—to form the epitaxial structure into a working device. The unique feature of modern devices is the reliance on the epitaxial formation of the material's structure. While several device formation techniques are discussed elsewhere in this volume, this chapter will concentrate on the evaluation of the epitaxial structures used in optical devices. The pre-device fabrication characterization of the epitaxial wafer is an essential part of this technology. The device performance in many ways is set at the moment of growth of the structure.

This chapter presents these characterization techniques in the context of the important film properties required in these structures. The techniques and procedures are not unique to any specific crystal growth technique and are generally used by all crystal growers and device engineers. The structural aspects and optical and electrical characterization of individual layers and multilayer structures are addressed. The correct physical structure alone is not sufficient for ensuring the formation of high-performance devices. These layers and structures must also possess the proper electrical and optical characteristics. The essential information required for the interpretation of device characteristics are (1) the physical properties of the film, such as composition and thickness of the individual layers, (2) doping and unintentional impurity incorporation, (3) optical properties such as luminescence and wavelength dependence, (4) interface structure and properties, and (5) the strain induced in an individual layer due to a lattice mismatch. The uniformity of these properties is critical to achieving a high yield from the fabricated devices and has received a great deal of attention in the pre-device fabrication characterization.

There are many complementary techniques which have become standard in the characterization of these structures. Typically, no one technique can unambiguously yield the desired information. Data from a variety of techniques must be correlated, and a self-consistent description should arise. The range of common techniques are given in Figure 2.3. The acronyms pictured in this figure are discussed in the following text. This figure indicates the multifaceted approach to materials and device characterization necessary in the development of optical device structures. These techniques are discussed below in the context of the information required from the device structure.

2.2 Growth Rate/Layer Thickness

In Situ Growth Monitors

The development of the epitaxial growth process for optical devices starts with the simple determination of the growth rate, composition, and morphology of a film. The growth rate within the epitaxial system, as well as the thickness of the individual layers, is required for one to interpret results obtained from device measurements. The thickness of specific layers is crucial in most device structures. Ideally, the growth rate would be measured during the actual layer growth or the real-time thickness of the growing layer monitored. Such "in situ" monitoring techniques are currently being developed. In the case of MBE, the growth temperature is typically monitored by an optical pyrometer. The pyrometer signal, however, oscillates during the growth of a heteroepitaxial layer due to optical interference of the infrared light passing through the sample.[2] As the heteroepitaxial layer grows, it forms an "anti-reflection" coating on the substrate, resulting in the oscillations of the measurement signal. Knowledge of the optical constants of the growing material combined with the oscillation period of the pyrometer signal can yield the growth rate or corresponding layer thickness.

The development of in situ probes applicable to both MOVPE and MBE growth systems is ongoing. These are generally optically based, such as with optical laser reflectivity at the visible or near-visible wavelengths. Oscillations in the reflected light intensity can again develop, as in the case of the optical pyrometer. Such reflectivity techniques can be implemented for real-time measurements of growth thickness while the device structure is being fabricated. The analysis of the intensity

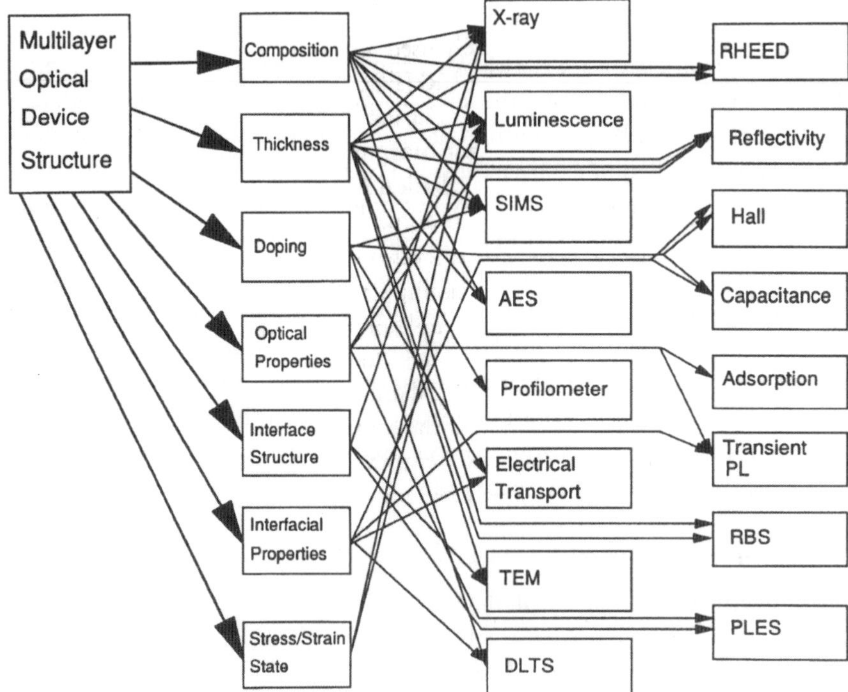

Figure 2.3 The characterization of the epitaxial wafer used in optical device fabrication requires a multifaceted approach to develop a full understanding of the materials structure. A particular measurement can yield information on several materials properties, increasing the value of that technique.

and variation of the reflected light is easily modeled using standard geometrical optics.[3] Control of the composition of a growing film can be accomplished by the real-time calculation of the reflectivity, comparison to the experimentally determined value, and the subsequent adjustment of the growth controlled parameters. Post-growth measurement of the layer thickness can also be determined optically through ellipsometry.

Newer optical probes, such as reflection difference spectroscopy (RDS) and spectral ellipsometry, are currently being developed as real-time in situ growth monitors. The RDS spectrum exhibits oscillations, under suitable conditions, that can be related to the growth of a single atomic layer of material. Spectral ellipsometry can be used to determine both composition and thickness, provided the optical constants of the growing materials are approximately known. Both these measurements,[4] as well as those described above, require direct line-of-sight access to the growth environment. Growth-related changes in the optical path, such as fogging of the optical ports by extraneous deposited material, must be controlled or taken into account during the measurement, limiting the applicability of these measurements.

Specific to the MBE-based growth techniques is reflected high energy electron diffraction (RHEED). RHEED is usually used to monitor the surface stoichiometry and structure of the growth surface.[5] RHEED has also been used as a tool for monitoring growth at the atomic level. During the layer-by-layer growth of the semiconductor, the intensity of the main Bragg reflection of the RHEED pattern oscillates in intensity. The period of oscillation has been shown to correspond to the growth of a single monolayer of the semiconductor. The physical origins of these oscillations are still a matter of discussion; however, RHEED oscillations provide an accurate, rapid means of obtaining growth-rate data. This technique is used to calibrate the growth rate when employing a specific source Knudsen cell. Its application to real-time analysis is limited, due to degradation of the materials properties under electron-beam exposure, damping of the oscillation with layer thickness during continuous growth, and the somewhat specialized growth conditions required to obtain the oscillations. Despite these limitations, RHEED oscillations can be used to rapidly calibrate growth rates just prior to growth of the actual device wafer.

Post-Growth Structural Analysis

The post-growth determination of the layer thickness is more commonly made by one of several usually destructive techniques. For thicker layers, the substrate-epilayer combination can be cleaved and the layers chemically stained. The chemical stain darkens or highlights layers, with the degree of staining dependent on the specific composition or doping. The cleaved sample edge is later inspected after staining by optical microscopy, typically using a technique such as Nomarski phase contrast, in order to determine the layer thickness. The thickness of very thin layers can also be determined through the inspection of the cleaved and stained edge by scanning electron microscopy (SEM).[6] In all these cases, the accuracy of the thickness measurement is dependent on a determination of the magnification of the instrument. Very thin layers can only be imaged by the time-consuming and expensive process of cross-sectional transmission electron microscopy (TEM). The TEM micrograph seen in Figure 2.4 was taken from an $Al_{0.3}Ga_{0.7}As/GaAs$ multilayer structure. The individual layers are clearly seen in the micrograph, allowing for the determination of the various layer thicknesses. The high magnification achievable in TEM[6] allows for almost atomic resolution on the TEM image. This process, while yielding detailed information, is not amenable to routine structure determination. The determination of the actual interface abruptness, that is, the spatial variation in composition, is very difficult to obtain from these micrographs, since the actual change in contrast is due to a variety of instrumental and sample-dependent factors.

A profilometer can also be used to mechanically measure the thickness of a layer. Stylus profilometers can accurately and rapidly determine physical step heights etched into a sample surface. Many profilometers can measure step edges with a resolution of a few nanometers. Wet and plasma-based etches are available for most

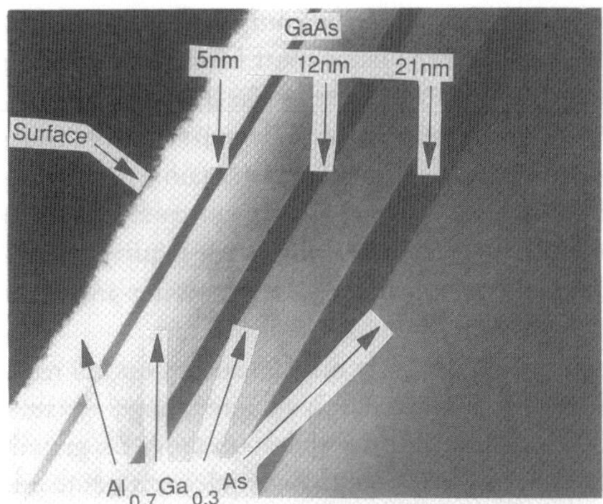

Figure 2.4   Transmission electron micrograph of an Al$_{0.7}$Ga$_{0.3}$As/GaAs multi-layer structure. The dark bands in the micrograph are thin layers of GaAs (5, 12, and 21 nm in thickness) between thicker layers of Al$_{0.7}$Ga$_{0.3}$As. The contrast in these types of micrographs allows for the determination of the thickness of the individual layers but does not yield exact compositional information.

alloy semiconductor systems which are highly compositionally selective. For example, Al$_x$Ga$_{1-x}$As, with $x > 0.50$, can be rapidly etched in HCL or HF solutions, while GaAs is relatively untouched. A wide variety of selective etches have been developed for all the standard ternary and quaternary optical materials. An overlying epitaxial layer can be selectively removed down to an underlying layer by masking an overlayer with photoresist or another etch-resistant material. The profilometer can be used to find the thickness of the overlayer once the etch mask is removed. The ease and convenience of this technique combined with the general availability of profilometers is responsible for its widespread use, despite the destruction of some of the sample area.

2.3 Composition Analysis

The composition of the material, in addition to layer thickness, determines a laser optical emission wavelength, the cutoff wavelength of a detector, and any property which is dependent on bandgap. The direct band edge of Al$_x$Ga$_{1-x}$As increases, for example, ~14 meV for $\Delta x = 0.01$. This change in composition corresponds to a wavelength shift of ~7 nm in the light emission for compositions near GaAs. In the lattice-matched ternary alloy systems, such as the Al$_x$Ga$_{1-x}$As/GaAs system, the band edge is determined only by the relative ratio of the two constituent binaries in the material, such as the AlAs/GaAs ratio in Al$_x$Ga$_{1-x}$As. More complex structures involving quaternary materials, such as In$_x$Ga$_{1-x}$As$_y$P$_{1-y}$, require control of multiple species. This implies that two ratios must be controlled in quaternary materials; for example, both the In/Ga and As/P ratios must be controlled in the growth of $In_xGa_{1-x}As_yP_{1-y}$. In non-lattice-matched systems, the bandgap is a function of the strain state of the layer. The strain state must be determined from other physical measurements. Composition control is required to maintain a desired lattice match between the epilayer and the underlying substrate. In epilayers which must be lattice-matched to a substrate, the degree of lattice match required is usually better than $\Delta a/a \lesssim 10^{-3}$, where a is the substrate lattice parameter and Δa is the difference in lattice between the epilayer and substrate parameters.

Layer composition can be determined by a variety of physical and optical techniques. The choice of techniques depends on the accuracy required. Auger electron spectroscopy[6] (AES) is a high-vacuum technique compatible with the MBE growth system. AES is based on the emission of Auger electrons under electron bombardment. Since the energy of an Auger electron is specific to a given element, the energy distribution of the Auger electrons can yield the composition of the sample.[7] The sensitivity of AES is usually on the order of ~0.5% which is insufficient for many applications. The addition of an ion gun, which removes the surface region by sputter erosion, to the AES system does allow for the depth profiling of the elemental composition in a multilayer sample. A related physical analysis technique is electron microprobe analysis. In this case, an incident electron beam generates X-ray emission from the constituent elements comprising the sample. The volume sampled by the electron probe is large. The electron beam penetrates into the sample, generating X rays over a depth of 0.5 to 1 µm, which limits this technique to the analysis of thick epitaxial layers. Electron microprobe characterization of thin layers demands extensive data analysis, typically requiring additional information concerning the sample geometry or composition.

High-resolution X-ray diffraction is a convenient, quantitative means of determining the structural properties of a multilayer structure. Commercial, user-friendly, double crystal X-ray diffractometers (DCXRD) along with diffraction simulation programs allow for the acquisition of an X-ray spectrum and its direct modeling. As a result, DCXRD has been developed into the premier technique for determining the composition of multicomponent semiconductors.[6] The measurement technique, when used on samples containing nearly or actually lattice-matched layers, is referred to as a rocking curve measurement or spectrum. Examples of a rocking curve from two separate single $Al_xGa_{1-x}As$ layers grown on GaAs are shown in Figure 2.5.[6] These spectra contain information concerning layer thickness and composition. The position of the Bragg reflection peaks from the epitaxial layer, $\Delta\Theta$, relative to the substrate peaks allows the determination in the difference in lattice parameter from the substrate. Since the substrate lattice parameter is known, the epilayer lattice parameter is determined and the composition deduced from known lattice constants. The data analysis for thick layers is well understood and simple to carry out. In Figure 2.5, the angular splitting between the substrate and $Al_xGa_{1-x}As$ peaks is small, less than 300", indicating a close lattice match

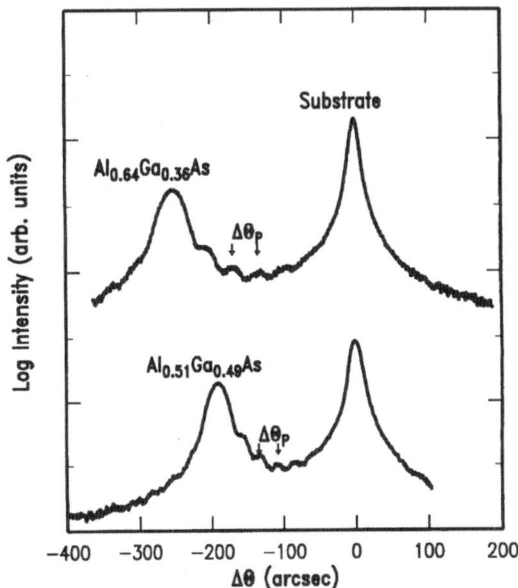

Figure 2.5 The double crystal X-ray rocking curve taken from two separate samples of Al$_{0.64}$Ga$_{0.36}$As and Al$_{0.51}$Ga$_{0.49}$As grown on GaAs. The close lattice match between the Al$_x$Ga$_{1-x}$As layers and GaAs results in pseudomorphic growth. The separation of the two principal narrow Bragg peaks yields compositional information while the additional oscillatory structure, $\Delta\Theta_p$, is related to the film thickness.

between GaAs and Al$_x$Ga$_{1-x}$As. The analysis of the data for such pseudomorphic layers also requires knowledge of the Poisson's ratio of the material, typically possessing a value of about 0.3. The determination of whether or not a layer is pseudomorphic can be ascertained, generally, from the width of the epilayer Bragg peak. Large peak widths for epitaxial layers are sometimes a strong indication of the appearance of extended defects arising from the relaxation of the pseudomorphic growth. In these single-overlayer structures, the DCXRD spectrum yields an oscillating intensity between the two principal Bragg reflections, one from the substrate and one from the overlayer, also shown in Figure 2.5. These oscillations, known as Pendelossing oscillations, have a period, $\Delta\Theta_p$, that is related to the single-layer thickness.[6] The DCXRD spectrum from superlattice structures can determine the period of the repeating unit and the average thickness of the individual layers, such as the Al$_x$Ga$_{1-x}$As or GaAs thickness in an alternating structure. As indicated, the DCXRD measurements on ternary materials can yield composition directly, since only the simple mixture of two binary compounds needs to be considered (e.g., InAs/GaAs in In$_x$Ga$_{1-x}$As). Additionally, all ternary and quaternary materials follow Vegard's law, simplifying the analysis. The determination of the exact composition of a quaternary material is more complex, requiring additional data such as the bandgap. The combined data of lattice parameter and bandgap allows the determination of the composition in closely lattice-matched systems.

DCXRD measurements and analysis can also be performed on multilayer structures in order to determine the details of the final structure. The DCXRD spectrum from a multilayer structure contains a great deal of fine structure, due to complex interference effects between X rays reflected from many layers, which can be compared to calculated spectra based on test or hypothetical structures. A typical DCXRD rocking curve from an $Al_xGa_{1-x}As/GaAs$ multilayer structure is shown in Figure 2.6, along with the result of a dynamic simulation of the rocking curve.[8] The structure is composed of a GaAs substrate, 445 nm layer of $Al_{0.33}Ga_{0.67}As$, 183 nm layer of GaAs, and a 460 nm surface layer of $Al_{0.33}Ga_{0.67}As$. The dynamic simulation, often performed using commercial software, can accurately model this multilayer sample and predict the many fine-structure features in the DCXRD spectrum. A comparison of the experimentally determined rocking curve to the calculated curve offers a means of characterizing a multilayered sample in a fairly accurate fashion, depending on the precision of the data and the details of the calculation. This nondestructive measurement can be performed over the surface of the wafer, determining uniformity in thickness and composition.

The composition of multicomponent materials can also be accurately obtained from photoluminescence (PL) measurements.[6] PL from an alloy, typically obtained at low temperatures of 77 K or below, can be obtained rapidly in a nondestructive

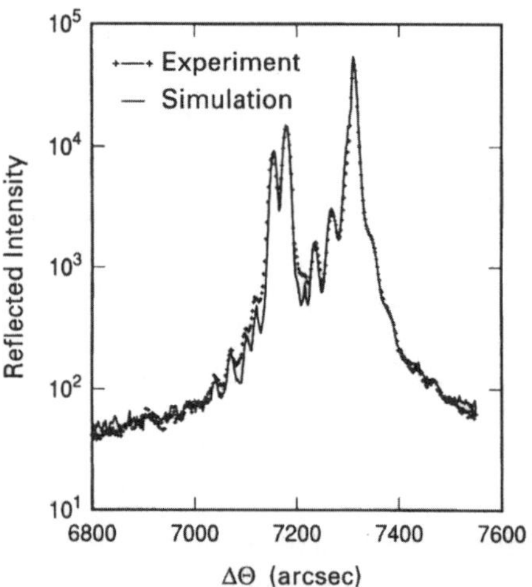

Figure 2.6   The DCXRD rocking curve obtained from a structure composed of a GaAs substrate, 445 nm layer of $Al_{l0.33}Ga_{0.67}As$, 183 nm layer of GaAs, and 460 nm surface layer of $Al_{0.33}Ga_{0.67}As$ is modeled using a dynamic simulation of the X-ray scattering from a multilayer structure. A comparison between the experimental data and the stimulation can be used to determine the composition and thickness from multilayer structures.

fashion. The band-edge luminescence spectra has well-defined features for most ternary and quaternary materials. PL spectra from $Al_xGa_{1-x}As$ layers grown on GaAs substrates is shown in Figure 2.7.[9] The highest energy peak is associated with either a band-to-band radiative transition (at higher measurement temperatures) or various excitonic and impurity-related transitions (at very low temperatures). The energy of these transitions are related to the band structure and bandgap of the material. The lower Al mole fraction $Al_xGa_{1-x}As$ samples shown in Figure 2.7 are direct bandgap materials, while the higher composition sample is an indirect gap semiconductor. The change from direct to indirect band structure is accompanied by a shift in the structure of the luminescence spectrum. Once identified, the band structure can be used to derive the sample composition. The composition is related to the bandgap provided the strain states of the materials are known. PL from pseudomorphic layers, particularly under a high degree of strain, exhibit predictable shifts in the bandgap from the relaxed or nonstrained case. The strain must be determined from other measurements, DCXRD in particular. The composition in nonstrained ternary alloy layers can be obtained from PL measurements through a comparison to known bandgap-versus-energy curves. Both DCXRD and luminescence maps of the composition over the wafer surface can be obtained. These combined measurements can yield an accurate view of the compositional uniformity over the substrate surface.

2.4 Impurity and Dopant Analysis

The concentration of impurity atoms within a layer is difficult to quantitatively determine due to the high sensitivity required in the depth profiling of trace atoms. Secondary ion mass spectrometry (SIMS) measurements can be a quantitative profiling technique with an extremely large dynamic range.[6, 10, 11] The sample surface

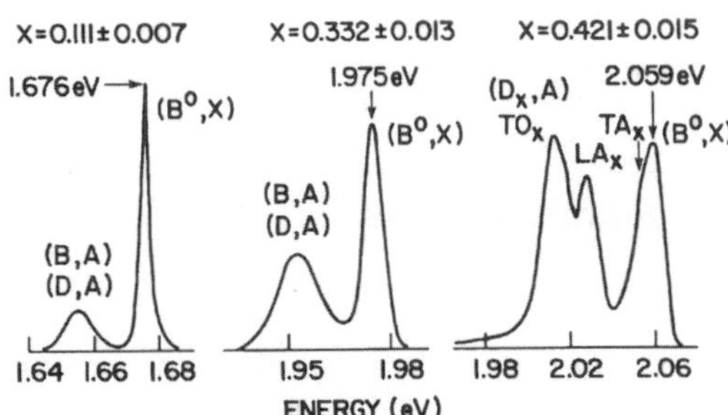

Figure 2.7   The PL spectrum from an alloy compound semiconductor can be used to determine the composition of the sample. The spectra shown here are obtained from high-purity $Al_xGa_{1-x}As$ samples grown by the MOVPE technique. The change in the spectral signature from the sample between $x = 0.33$ and $x = 0.42$ indicates the shift from a direct to indirect band structure, which occurs at a composition of about $Al_{0.37}Ga_{0.63}As$.

in the SIMS technique is sputtered, and those sputtered atoms which leave the surface as ions, referred to as secondary ions, are collected in a high-resolution mass spectrometer. The intensity of the signal is related to the composition of the sample. The mass spectrometer has a very wide dynamic range, which allows for the measurement of major constituents as well as trace impurities. The SIMS spectra in Figure 2.8 contain depth profiles of both ^{12}C and ^{13}C as impurities in GaAs.[12] The carbon incorporation was stepwise altered during growth, yielding a steplike profile to the impurity concentration. A dynamic range of about 4 decades was obtained in the ^{13}C spectrum in this figure. The limits to the overall dynamic range of the SIMS measurement extend all the way to the pure elements, yielding an overall range of $\sim 10^8$. The absolute sensitivity limits are set by the instrument and the sample, but sensitivities on the order of 1 ppm or less can be achieved for impurities in most semiconductors. Despite this seemingly straightforward procedure, quantitative SIMS measurements can be difficult to obtain due to experimental complications which must be keep in mind. The yield of secondary ions is dependent on the host matrix containing the impurity. These "matrix" effects necessitate the use of an internal calibration standard for the quantitative measurement of impurity levels. This is most easily achieved through the use of an ion implant of the sought impurity into the surface of the sample of interest. The measured signal from the implanted region can be compared to the known dose of the implant, affording a ready calibration standard free of matrix effects. A peak from ion-implanted ^{13}C superimposed with the ^{12}C peak is seen in the near-surface region of Figure 2.8, providing an internal calibration of the carbon concentration. The wide dynamic range and quantitative nature of the SIMS technique has expanded its role in the characterization of semiconductor materials.

SIMS is limited in the depth resolution that can be typically achieved due to sputter-induced mixing in the sample. Typically, for a large step change in an impurity concentration in depth in the sample, the SIMS measurement is limited to a decade drop in measured concentration for about a 10 nm sputter depth change. This places a limit on the ability of SIMS to determine very rapid changes in composition and impurity concentration.

2.5 Electrical Properties in Optical Structures

The electrical properties carrier concentration mobility and resistivity are generally controlled by intentional impurities or dopants introduced during growth. In most optical devices, highly controlled doping of the device structure is not required. This is in sharp contrast to the high degree of doping control and uniformity required in electronic charge control devices such as field effect transistors (FETs). There are two standard techniques used in the measurement of doping and resistivity: Hall and capacitance–voltage (C–V) measurements. Hall measurements are used to obtain the free carrier concentration and mobility. This measurement is generally destructive, requiring test structure formation and metallization. The Hall

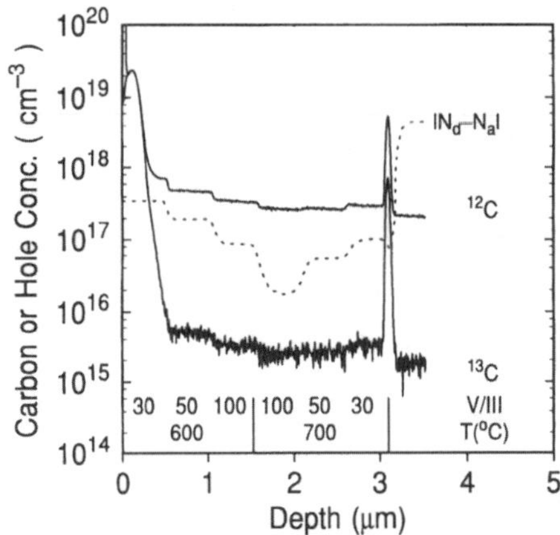

Figure 2.8 The SIMS profiles of intentionally carbon-doped GaAs reveal the sensitivity and wide dynamic range of this technique which is necessary in the analysis of very minor constituents such as dopants. The near-surface region contains an ion-implanted carbon peak used as an internal calibration for the SIMS measurement.

coefficient determined from the measurements on uniformly doped samples can be directly related to the depth of integrated carrier concentration, or the sheet concentration of carriers on single-layer samples. The conversion to a volumetric concentration requires a knowledge of the surface depletion region formed by the pinning of the Fermi level due to the high density of surface states found at most compound semiconductor surfaces and many interfaces. For example, the surface depletion width is ~0.1 μm for GaAs doped to an electron concentration of 10^{17} cm^{-3}. The electrical thickness used in calculating the bulk carrier concentration would therefore be smaller than the metallurgical thickness by these depletion regions; that is, the electrically active thickness of a 1 μm layer of GaAs:$n = 10^{17}$ cm^{-3} would be 0.8 μm, accounting for surface and interface depletion regions. Hall measurements taken from multilayer structures, where several layers are conductive, are difficult to interpret, since an aggregate sheet carrier concentration is obtained as a result of the coupled and parallel conduction between the multilayers.

Hall measurements also yield the mobility of the layer. The purity of undoped or lightly doped materials used in the active regions of lasers or that are required in photodetectors can be evaluated by noting the 77 K mobility. The purity of undoped materials is often cited in these terms. The loss of doping control, particularly at low doping concentration, can be analyzed by the low-temperature mobility. The mobility value at 77 K is dominated by scattering from ionized impurities. The presence of a low mobility value, for a given carrier concentration, indicates the presence of compensating impurities or defects. The higher the concentration of such impurities, such as unintentional carbon in n-type GaAs, the lower the mobility. Tabulated values of mobility versus carrier concentration at various levels of

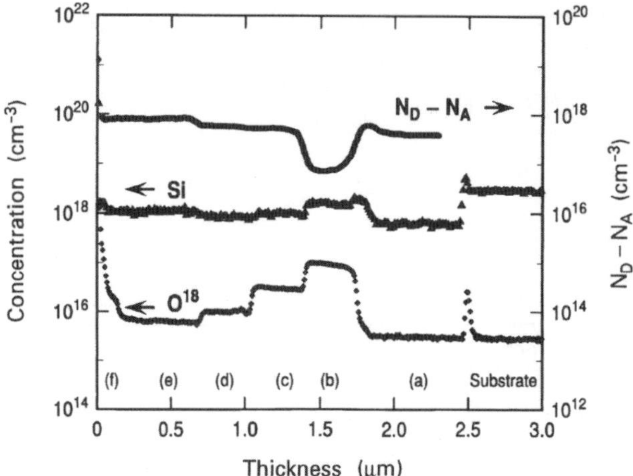

Figure 2.9 The electrochemical capacitance profile of a GaAs sample co-doped with Si and ^{18}O. A comparison of the carrier concentration, N_d–N_a, to the SIMS profiles of Si and ^{18}O indicates that the carrier concentration attributed to the Si doping is compensated by the oxygen deep level. The carrier concentration is roughly equal to the difference between the Si and ^{18}O. Such combined SIMS and capacitance measurements allow for the diagnosis of contamination-related doping problems.

compensating impurities allow for the quantitative evaluation of the concentration of these defects,[13] at least in higher-purity films (e.g., $<10^{16}$ cm^{-3} in GaAs).

Capacitance measurements can be readily performed on the exposed surface region of the semiconductor. The differential C–V characteristic can yield the carrier concentration profile of the near-surface region. These measurements can be combined with electrochemical etching, which allows for the profiling of the carrier concentration through a complex multilayer structure. The changes in carrier concentration in the depth of the sample can be easily obtained through such an electrochemical profiling technique. Figure 2.9 has an electrochemically derived carrier profile of GaAs co-doped with Si and ^{18}O. The oxygen compensates the Si dopant, leading to a reduction in the carrier concentration from the expected value, which would be that of the Si concentration. The Si and ^{18}O profiles were determined by a SIMS measurement. The combination of Hall, capacitance, and some physical profiling measurements can yield the complete electrical profile of the structure, as well as point to any anomalous features.

Hall and C–V measurements should, in principle, agree in the case where there are only simple donors or acceptors in the materials. Infrared lasers and other optical devices rely, however, on ternary materials in which the typical donors, such as Si and Al$_x$Ga$_{1-x}$As, form several levels. These donors form so-called DX centers. DX centers are associated with persistent photoconductivity effects and carrier freeze-out for alloys within certain composition regions. Since DX centers are an

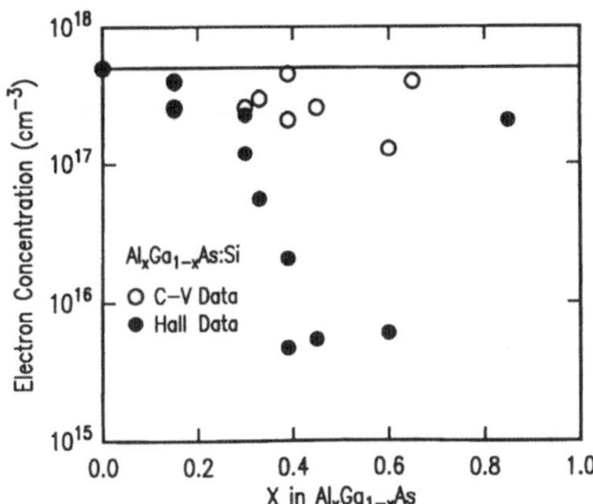

Figure 2.10 The free electron concentration determined by Hall effect measurements and the net donor concentration determined by capacitance measurements differ considerably within the immediate composition range of $Al_xGa_{1-x}As$:Si.[13] This behavior is due to DX centers formed by the donors in the $Al_xGa_{1-x}As$ layer.

inherent property of donors in many alloys, they must be understood in order to develop an understanding of many doping processes. A recent review of these donors has been published describing their properties.[14] These centers can lead to a large discrepancy between C–V and Hall measurements. Room temperature C–V measurements in these DX center-dominated materials yield carrier concentrations indicative of the net active donor concentration. The Hall measurement, however, can yield a carrier concentration orders of magnitude lower than the C–V derived value. The Hall measurements yield the concentration of carriers ionized from the donors, while C–V measurements yield the total donor concentration. This is shown for the case of $Al_xGa_{1-x}As$:Si as a function of alloy composition in Figure 2.10.[15] In practical terms, the carrier concentration in several layers can be quite low, yielding high resistivity values for those layers, despite the large concentration of electrically active dopants. Since the DX state is an intrinsic property of the donors in these materials, there is no available means of altering this persistent and troublesome behavior. Knowledge of the electrical behavior of these states is necessary for the correct interpretation of the electrical measurements and for proper device design.

Deep electrically active levels within the bandgap of the active region of a laser structure or LED can prevent the efficient generation of light. Oxygen and transition metals, for example, are known to kill or greatly diminish radiative lifetimes in semiconductors, strongly reducing their luminescence efficiency. The presence, concentration, and properties of such deep levels can be determined within actual device structures, in addition to simple single layers, through deep level transient spectroscopy[16] (DLTS). DLTS is a time-resolved capacitance measurement which

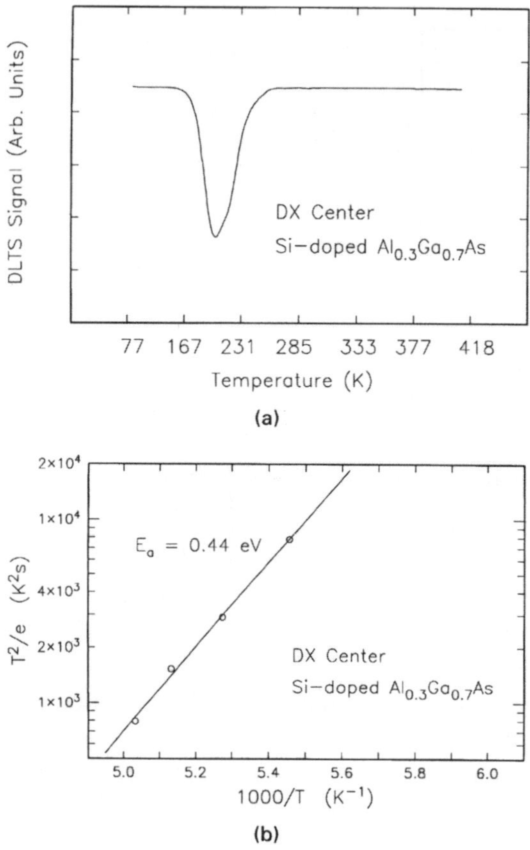

Figure 2.11 DLTS can yield the energy position of a deep state within the bandgap of the semiconductor. (*a*) The DLTS signal is related to a transient change in capacitance within the structure in response to a voltage pulse, shown here for Si DX-related centers in Al$_{0.3}$Ga$_{0.7}$As. (*b*) This signal is then related through an Arrhenius plot to the emission energy of the deep state, which can lead to the identification of a deleterious impurity or defect.

can reveal the presence of electrically active deep states. The simple structures required for a DLTS measurement, often only a Schottky diode, receive a short voltage pulse over a set dc bias. The traps can capture (emit) carriers during this pulse, and the relaxation of the capacitance to its quiescent value is monitored. Since emission (capture) is a thermally activated process, the rate of this process can be monitored—that is, the change in capacitance with time—as a function of measurement temperature, as shown in Figure 2.11*a*. The Arrhenius plot of the maximum emission rate, approximately given by the maximum change in capacitance, as a function of temperature under a given set of experimental parameters can yield the emission energy, as given in Figure 2.11*b*. This emission energy is associated with the position of the deep state within the bandgap, and the concentration of a deep state is deduced from the amplitude of the DLTS signal. Many of these emission energies have been related to a specific chemical or structural impurity, allowing for the identification of a particular deleterious defect. It is not always clear, however, that the detected deep level is solely responsible for the decreased luminescence. Correlations between the concentration of deep levels, their energy position in the bandgap, and luminescence measurements can identify a key defect as the problem within a material.

2.6 Optical Properties in Single and Multilayer Structures

Light-emitting structures require both a high luminescence efficiency and a low concentration of below-bandgap states. Nonradiative deep states provide a competing recombination pathway, diminishing the quantum efficiency of the optical structure. The optical properties can be qualitatively evaluated from photoluminescence, as mentioned above, and cathodoluminescence (CL) measurements. The difference between these measurement techniques is in the source of excess carriers: PL derives carriers from the illumination of the sample with above-bandgap (direct edge for efficient generation) light, while CL utilizes electron-beam excitation.[6] CL does allow for the local excitation of individual parts of a multilayer structure through the selective exposure of the layers to a finely focused and directed electron beam. These measurements are typically incorporated into a scanning electron microscope (SEM), allowing for surface imaging as well as luminescence measurements.

PL or CL spectra yield a variety of data useful in the compositional and physical analysis of optical device structures. The luminescence spectra obtained from single-layer structures possess several spectral features which typically can be easily interpreted, providing structural, purity, and compositional information. Compositional information can be obtained from the band-edge energy, if the luminescence feature can be identified, as discussed in a previous section. The ease of performing PL measurements, particularly at room temperature, enables the rapid acquisition of band-edge energy as a function of position over the wafer surface. The PL mapping of a wafer surface can effectively display the compositional uniformity of the epitaxial layer. These PL maps provide a means of studying irregularities within the growth system, which result in unexpected variations in compositional uniformity over the surface. PL maps of the luminescence intensity can also be used to identify regions containing extended or point defects, which result in diminished PL intensity. Strain relaxation in pseudomorphic layers can often be identified in this way.

Luminescence measurements have been applied to determining the thickness in quantum-sized (< 20 nm) layers in multilayer heterostructures. These luminescence measurements are often the easiest means to determine the quality and physical properties of thin layer structures. The quantum confinement induced by the diminished layer thickness causes a blue-shift in the emitted light from the structure, as indicated in Figure 2.12.[17] The lowest energy level in the smaller bandgap or well material (GaAs) is shifted to the blue by an amount approximately dependent on layer thickness, $\Delta E_{\text{blue}} \propto 1/a^2$. The specific amount of blue shift can be calculated using an envelop function approximation for a given materials system. Figure 2.12 contains both a TEM micrograph of the physical structure as well as a schematic of the band structure through the multilayer $Al_xGa_{1-x}As$/GaAs structure. This figure indicates the increasing shift of the radiation with diminishing well thickness. The luminescence spectral position can yield layer thickness information, provided assumptions such as abrupt compositional profiles are met within the structure. The luminescence linewidth can also provide information on the uniformity of the thin layer structure, as well as the purity level. The linewidth is often cited as a figure of merit in the formation of quantum structures, with narrower linewidths generally indicating higher purity and enhanced, smoother, or more uniform heterointerfaces. The luminescence linewidth is affected by a variety of factors. The linewidth is broadened by elevated levels of impurities in the well and the barrier materials. Thickness variations also can lead to line broadening, as indicated schematically in Figure 2.13. Thickness variation on a length scale greater than the exciton diameter (about 20 nm in GaAs) will result in light emission from wells of varying thickness, as described in the figure caption. Since these adjacent well regions usually differ by only a few atomic layers, the energy separation between such peaks, close in energy, is not well resolved, and a broadened luminescence line results. The analysis of luminescence position and linewidth is therefore used to study the structure of the thin layer sample and can be used to diagnose difficulties in impurity incorporation, junction placement (through the appearance of impurities in the well material), and interfacial structure.

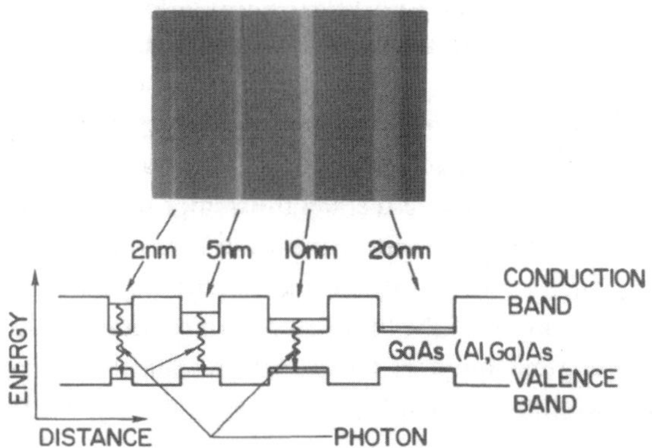

Figure 2.12   The electronic structure of a multi-quantum well structure is schematically shown compared to its physical structure given by the TEM micrograph.[15] The thin layer structures yield a shift in the energy or wavelength of the emitted light. The luminescence spectrum of the multilayer structure can be used to ascertain the physical dimension of the quantum-well layer.

2.7 Interface Properties in Multilayer Structures

The controlled formation of doping and compositional interfaces is crucial to the fabrication of high-performance light-emitting devices. As indicated above, several structural probes serve to determine the physical structure of the multilayer film. The performance of a device can be determined by factors other than the physical structure alone. Interface properties lie in three main areas: junction placement, compositional grading, and electrical behavior of the interface.

Junction placement is affected by a variety of growth and post-growth processing-related factors. The accurate placement of the dopant atoms within the device structure can be complicated by the appearance of dopant transients within the structure. This is shown in Figure 2.14 for the case of GaAs:Mg grown by MOVPE using $(C_5H_5)_2Mg$ or Cp_2Mg as the dopant source. The dopant source was abruptly turned on at a depth of about 1.3 μm and turned off at a depth of 0.9 μm. While the injection of the source into the growth reactor was a step function, the incorporation of dopant forms long transients related to the interaction of the dopant with the internal surfaces of the reactor. Cp_2Mg exhibits particularly long transients in the growth profile, with other dopant sources forming gready reduced or no transients. These transients place dopant atoms far from the expected depth within the structure and may lead to the misplacement of the electrical junction. MBE growth often exhibits similar doping transients, particularly with Be and Sn dopants. Be and Sn tend to segregate to the growing surface, riding on the growth front and slowly incorporating into the layer. These transients are often identified through SIMS profiling or from unexpected electrical characteristics

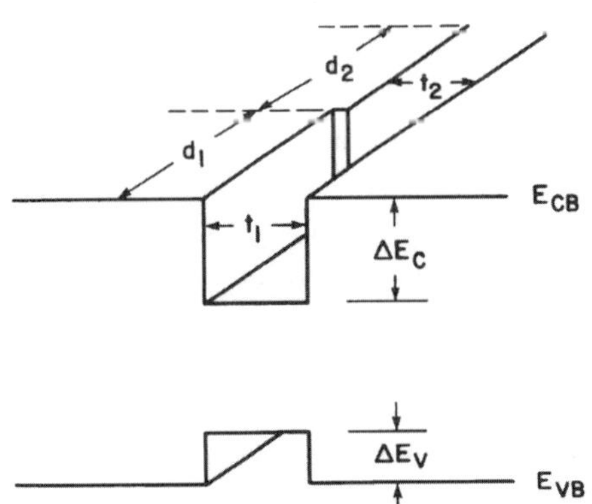

Figure 2.13   The luminescence linewidth is broadened by a variety of factors, including elevated impurity levels and well-thickness variations. The quantum confinement is created by the band edge offsets, ΔE_C and ΔE_V. If the excition diameter, d_{ex}, is smaller than the typical lateral dimension of the thickness variation, d_1 or d_2, the luminescence peak will be composed of superimposed light emitted from wells of differeing thickness, t_1 and t_2.

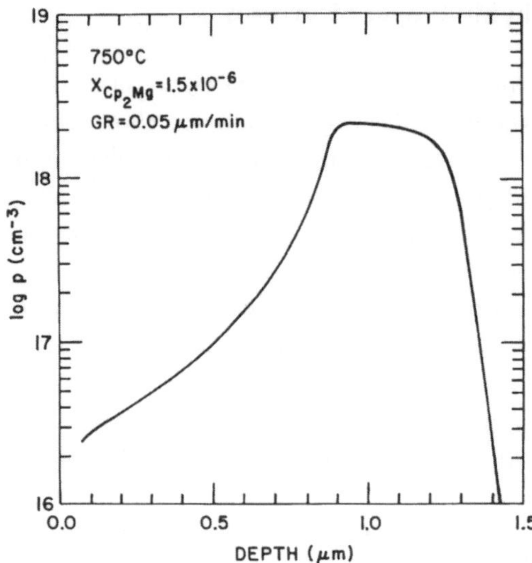

Figure 2.14 The growth of doped structures is complicated by a variety of growth-related factors. Junction placement, critical in heterojunction devices, depends on the details of the doping process. The interaction of the dopant with the growth system can lead to a slow "turn on" and "turn off" of the dopant profile, despite abrupt changes in the introduction of the dopant into the system. This is shown here for Cp_2Mg-based Mg doping of GaAs by MOVPE. The dopant profile decreases slowly over several hundred nanometers, limiting the utility of this particular dopant.

of the device, such as an anomalous turn-on voltage to a p–n junction within the structure.

The motion of dopants also occurs in post-growth thermal processing, as in the formation of ohmic contacts or ion-implantation annealing. At low concentrations, dopant impurities are often characterized by a concentration-independent diffusion coefficient. Many of the common impurities used in III–V device formation, however, display a very pronounced concentration dependence to the dopant diffusion. Si, Zn, Be, Sn, and other dopant elements exhibit a strong concentration-dependent diffusion at the dopant levels typically found within many device structures. High dopant concentrations can therefore lead to the rapid formation of a sharp diffusion front.[18] The final diffusion profile is dependent on the composition and placement of the dopants within the structure. Sn and Zn are particularly fast diffusers, especially when placed in the near-surface region of the structure where there is a ready source of vacant lattice sites. The microscopic diffusion mechanism has been extensively studied, as discussed by Tuck.[17] Fast diffusion is used to advantage in several applications, as in the impurity-induced disordering of super-lattice structures.[19] Analysis of such profiles is accomplished through SIMS measurements and, in the case of impurity-induced disordering, by TEM micrographs.

          III–V COMPOUND SEMICONDUCTOR FILMS . . .   Chapter 2

Compositional grading at heterointerfaces is difficult to quantitatively assess. Grading on the scale of a few nanometers can detract from the performance of a quantum-based device structure. Typically, TEM micrographs have been employed to determine the compositional grading. Sharp contrast between two layers in the TEM image is often used to assign a measure of interfacial abruptness to a hetero-interface. TEM images, however, can be difficult to unambiguously interpret due to the effects of the imaging conditions on the image formation. High-precision DCXRD measurements can be used on suitable structures to determine the appearance of appreciable interface grading. These physical measurements, when used in conjunction with luminescence measurements, can provide a more accurate picture of the compositional profile on the scale of a few nanometers. In particular, photo-luminescence excitation spectroscopy (PLES), adsorption measurements, and modulation spectroscopies, such as photoreflectance and electroreflectance, can yield information on the position of the higher-lying quantum states within quantum-confined structures. Comparison of these states to calculated levels based on a model compositional profile within the well can be used to estimate the compositional profile across the interface.

The electrical and optical properties of the heterointerface are directly accessible to measurement. The optical properties of the interface are affected by the presence of mid-gap recombination centers which increase the interfacial recombination velocity. A high interfacial recombination velocity serves as a sink for minority carriers, for example, injected into a quantum-well light emitter, reducing the light emission efficiency. The presence of such states can be probed by DLTS on suitable structures. Time-resolved luminescence measurements can be interpreted to yield an interfacial recombination velocity in heterostructure measurements.

2.8 Summary

The primary determinate of the performance of an optical device is the detailed characteristics of the initial epitaxial structure. This chapter summarizes the chief properties of the epitaxial structure in terms of its structural, compositional, electrical, and optical behavior. The means of determining these properties within single and multilayer structures are introduced and discussed. These characterization techniques are not particular to a specific growth technology, but are generally applicable to all epitaxial structures.

Acknowledgments

The author wishes to thank P. Mooney for providing Figure 2.11 and for discussions associated with the DX center. Joan Redwing is also acknowledged for her careful reading of the manuscript. The author would also like to acknowledge discussions with R. Potemski, M. A. Tischler, T. N. Theis, and G. Scilla.

References

1 *Semiconductors and Semimetals*, Vols. 32 and 33. (T. P. Pearsall, Vol. Ed., and R. K. Williardson and A. C. Beer, Series Eds.) Academic Press, San Diego, 1991. A general review of strained layer materials.

2 A. J. Springthrope and A. Majeed. *J. Vac. Sci. Technol.* **8**, 266, 1990.

3 O. S. Heavens. *Optical Properties of Thin Films*. Dover, New York, 1965.

4 D. E. Aspnes, R. Bhat, E. Colas, L. T. Florez, S. Gregory, J. P. Harbison, I. Kamiya, W. E. Quinn, S. A. Schwartz, H. Tanaka, and M. Wassermeier. In "Atomic Layer Deposition and Processing." (T. F. Kuech, P. D. Dapkus, and Y. Aoyagi, Eds.) *Proceedings*. Materials Research Society, Pittsburgh, 1991.

5 *Reflection High-Energy Electron Diffraction and Reflection Imaging of Surfaces*. (P. K. Larsen and P. J. Dobson, Eds.) Plenum, New York, 1988. For recent reviews and complete listing of references.

6 *Encyclopedia of Materials Characterization*. (C. R. Brundle, C. A. Evans, Jr., and S. Wilson) Butterworth-Heinermann, Boston, 1992.

7 *Methods of Surface Analysis*. (J. M. Walls, Ed.) Cambridge University Press, New York, 1988, chapt. 4.

8 M. S. Goorsky and T. F. Kuech. *Proceedings*. (T. C. Huang, P. J. Cohen, and D. J. Eaglesham, Eds.) Symposium on Advances in Surface and Thin Film Diffraction, Materials Research Society, Pittsburgh, 1991, p. 237.

9 T. F. Kuech, D. J. Wolford, R. Potemski, J. A. Bradley, K. H. Kelleher, D. Yan, J. P. Farrell, P. M. S. Lesser, and F. H. Pollak. *Appl. Phys. Lett.* **51**, 505, 1987.

10 *Reflection High-Energy Electron Diffraction and Reflection Imaging of Surfaces*. (P. K. Larsen and P. J. Dobson, Eds.) Plenum, New York, 1988, chapt. 7.

11 *Methods of Surface Analysis*. (A. W. Czanderna, Ed.) Elsevier Scientific, Amsterdam, 1975, chapts. 6 and 7.

12 N. I. Buchan, T. F. Kuech, G. Scilla, F. Cardone, and R. M. Potemski. *J. Crystal Growth*. **110**, 405, 1991.

13 D. A. Anderson and N. Apsley. *Semicond. Sci. Technol.* **1**, 187, 1986.

14 P. M. Mooney. *J. Appl. Phys.* **67**, R1, 1990.

15 T. F. Kuech, E. Veuhoff, and B. S. Meyerson. *J. Crystal Growth*. **68**, 48, 1984.

16   G. L. Miller, D. V. Lang, and L. C. Kimerling. *Ann. Rev. Mater. Sci.* **7**, 377, 1977.

17   T. F. Kuech and K. F. Jensen. In *Thin Film Processes II.* (J. L. Vossen and W. Kern, Eds.) Academic Press, Orlando, 1991, chapt. III-2.

18   B. Tuck. *Atomic Diffusion in III–V Semiconductors.* Adam Hilger, Bristol, 1988.

19   J. W. Lee and W. D. Laidig. *J. Electron. Mater.* **13**, 147, 1984.

Contacts

T. SANDS and S. A. SCHWARZ

Contents

3.1 Introduction

Metal-semiconductor interfaces have for several decades been the subject of intense scrutiny. Atomically abrupt uncontaminated interfaces exhibit a remarkable range of properties which continue to defy unified theoretical treatment. Practical compound semiconductor contacts are seldom abrupt or uncontaminated and, as such, provide formidable challenges to characterization and understanding. As compound semiconductor device dimensions are reduced, increasingly stringent demands are placed on the resident metallic contacts. An abrupt interface is a highly desirable feature in present day contacts and offers the advantage of being amenable to experimental characterization by an array of powerful high-resolution techniques. In this manner, fundamental studies of the metal-compound semiconductor interface may have practical implications.

Not surprisingly, studies of metal-semiconductor contacts have fostered significant developments and refinements in experimental techniques. The primary experimental probes of contact behavior are surveyed in this chapter, with emphasis on recent advances in techniques and their consequent impact on the fabrication and understanding of Schottky and ohmic contacts. A valuable overview of this field is contained in the monograph of Rhoderick and Williams.[1] A wide array of experiments, prior to 1985, are concisely summarized in the review of Palmstrøm and Morgan.[2] The current understanding of Schottky barrier formation

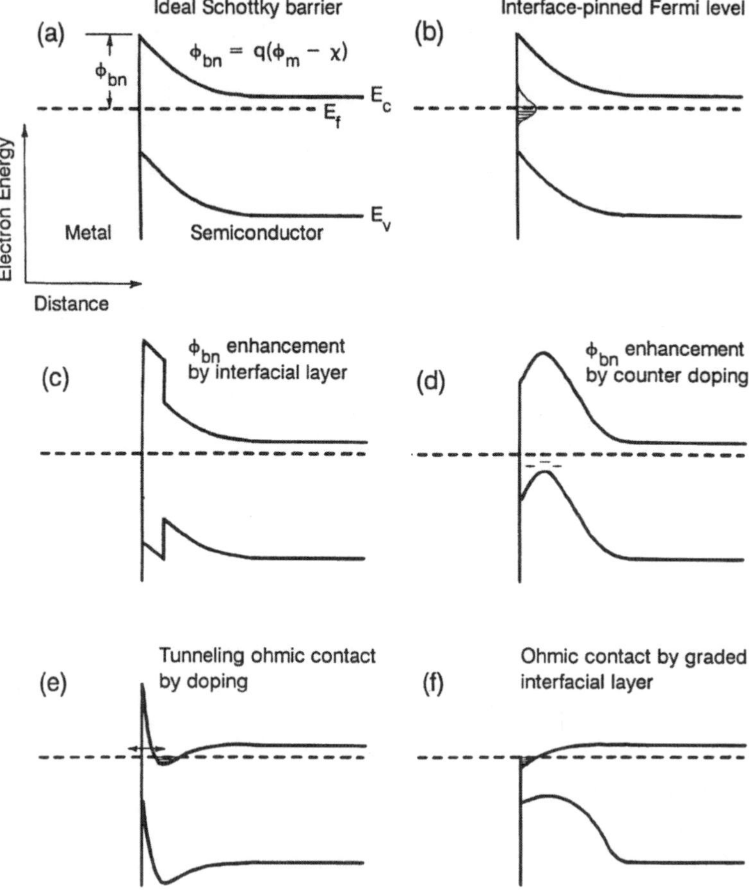

Figure 3.1 Schematic band diagrams for metal/*n*-type semiconductor interfaces: (*a*) ideal Schottky barrier (work function model); (*b*) rectifying contact formed by Fermi level pinned by mid-gap interface states; (*c*) barrier height modified by wide-gap interfacial layer; (*d*) rectifying contact based on heterotype "camel" diode model; (*e*) tunneling ohmic contact resulting from a degenerately doped interfacial layer; and (*f*) ohmic contact fabricated with graded bandgap approach to minimize effect of conduction band discontinuity.

is described by Mönch.[3] Ohmic contact technology has been reviewed recently by Sands[4] and Murakami.[5] Past, present, and future developments in the field may be monitored in the annual proceedings of the Physics and Chemistry of Semiconductor Interfaces (PCSI) conference and the American Vacuum Society (AVS) symposia, which are published in the *Journal of Vacuum Science and Technology*.

An important goal of characterization is the determination of the conduction mechanism, or applicable band diagram, for a specific metallic contact. Six plausible band diagrams for a metal-semiconductor interface are illustrated in Figure 3.1. Figure 3.1*a* shows an ideal Schottky barrier, with the barrier height, ϕ_b, determined by the difference between the metal work function and the semiconductor electron affinity. Charge is transferred from the metal to the bulk of the semiconductor to obtain the equilibrium alignment of Fermi levels. Alternatively, charge may be transferred to interfacial states within the semiconductor bandgap, as shown in Figure 3.1b. The Fermi level is then pinned at the position of the dominating gap state. Numerous experimental studies of Fermi level pinning fueled a vigorous debate over the origin of these gap states throughout the 1980s, the primary candidates being metal induced gap states (MIGS) and intrinsic semiconductor defects.[1] Recent observations of unpinned Schottky barriers and surface photovoltage effects have reinvigorated this debate, as described below. Figure 3.1c demonstrates the influence of a wide bandgap interfacial layer, several nanometers or less in thickness so as to be fully depleted. Strong modification of the effective barrier height may also be induced by a heterotype or homotype junction near the metal-semiconductor interface, as shown in Figures 3.1d and 3.1e. If this region is extrinsically doped n-type, the Fermi level moves into the conduction band of the semiconductor (Figure 3.1e). The resultant potential barrier is sufficiently thin to allow charge flow by tunneling, rather than thermionic emission, and the contact is rendered ohmic. In Figure 3.1f, a graded bandgap region at the interface also facilitates ohmic conduction. These band diagrams apply to uniformly abrupt two-dimensional contacts. Three-dimensional asperities may result from metallic spiking, intermetallic phase formation, defect propagation, evaporation through voids, and so on, enabling conduction at localized "hot spots." Furthermore, highly defective interfacial layers can lead to more varied conduction mechanisms.

Practical contacts are constrained by a number of requirements such as process compatibility, operational stability, bondability, and uniformity. As contact dimensions and operating voltages are scaled downward, it becomes increasingly difficult to maintain uniform barrier heights or contact resistances throughout an integrated circuit, to avoid interaction of the contact with buried epitaxial layers, or to maintain the voltage drop at the ohmic contacts at a relatively negligible value. One example of a shallow, morphologically uniform ohmic contact which addresses these diverse constraints is shown schematically in Figure 3.2. The PdGe/GaAs contact[6] is a non-alloyed contact, that is, it is not necessary to form a liquid interfacial phase at the contact formation temperature. In the as-deposited structure of Figure 3.2a, the interfacial Pd layer permeates the native oxide and reacts with the GaAs to form a 6 nm thick topotactic film of the Pd_4GaAs phase. At about 250 °C, the Pd and Ge overlayer interact to form PdGe. The Pd_4GaAs layer decomposes, resulting in the epitaxial regrowth of GaAs, which may be heavily doped with Ge. Figure 3.2 summarizes a large number of experiments probing interfacial morphology, phase formation, process dependence, and electrical behavior. Several of these studies are described below as illustrative examples.

This chapter is organized in parallel with the process flow, from in situ characterization of the starting semiconductor surface to analysis of the early stages of metal deposition, to characterization of thick unpatterned contact metallizations,

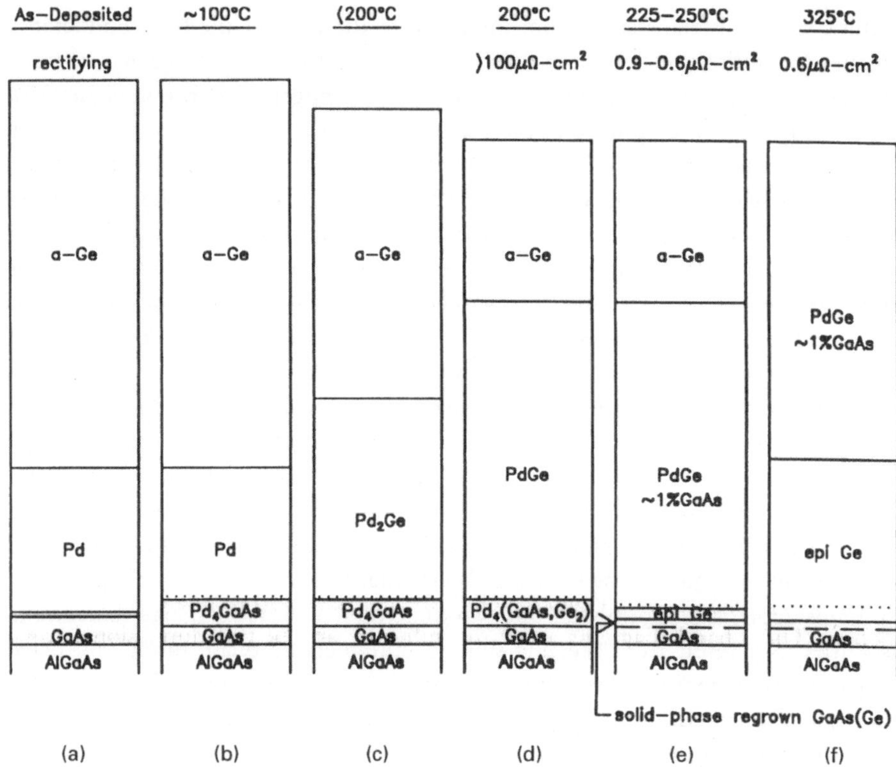

Figure 3.2 Sequence of reactions in the Ge/Pd contact to *n*-GaAs culminating in ohmic behavior: (*a*) as-deposited layer sequence; (*b*) initial reaction at ~100 °C to form Pd₄GaAs; (*c*) reaction at 100–200 °C to form Pd₂Ge; (*d*) continued reaction to consume Ge and form PdGe; (*e*) regrowth of GaAs (doped with Ge) resulting in ohmic behavior; and (*f*) crystallization of remaining Ge on GaAs. The dotted line indicates the original Pd/GaAs interface, marked in the early stages of reaction by residual native oxide and, in the later stages of reaction, by voids. (After Marshall et al.[6])

and finally, to methods of evaluating the properties of patterned contacts for device applications. (Descriptions of the analytical techniques used for materials characterization and referred to in this chapter can be found in the lead volume of this series, *Encyclopedia of Materials Characterization*. In addition, one-page summaries are collected in the appendix of this volume.)

3.2 In Situ Probes

Surface Preparation and Characterization

The starting semiconductor surface, even if uncontaminated and atomically abrupt, can profoundly affect the properties of the Schottky contact by virtue of its orientation, step or kink density, or surface reconstruction. This is exemplified by a recent study by Palmstrøm et al.[7] in which the Schottky barrier height for epitaxial metallic ErAs on GaAs varies strongly with surface orientation. In numerous UHV studies, GaAs is cleaved in situ, exposing a (110) surface that is unpinned, as evidenced by photoemission measurements. In semiconductor growth chambers, the (100) surface, which may be terminated by either Ga or As atoms exclusively, is typically employed. The native oxide on this surface is removed in situ by heating, typically above 580 °C in an arsenic flux.[8] Contamination in UHV is indicated by the quality of the low-energy electron diffraction (LEED) or reflected high energy electron diffraction (RHEED) pattern, which also indicates the surface reconstruction. Auger electron spectroscopy (AES) is also typically employed as a more sensitive and informative probe of contamination. The remarkable dependence of the (100) surface reconstruction on substrate temperature and Ga and As overpressure was elucidated in the RHEED study of Donner et al.[8] Al nucleation patterns on these surfaces were also examined. Recent experimental advances allow the surface layer stoichiometry to be monitored in situ by RHEED[9] or ellipsometry.[10] GaAs(100) surfaces prepared by MBE can be capped with As and transferred through air to a UHV analytical chamber. The As cap is then thermally desorbed, revealing a relatively high quality surface.[11] Unpinned surfaces have recently been prepared in air by chemical treatments such as S or Se passivation or photowashing.[12, 13] The surface is then covered by roughly a monolayer of H, S, or Se.

Initial Metal Deposition

The initial deposition of metal atoms onto the semiconductor surface has been studied extensively by photoemission spectroscopy. The electron energy distribution curves (EDCs) in these experiments convey several important pieces of information.[1] The shape of the curve reflects the near-surface density-of-states (distorted by the background of secondary electrons and by variations in sensitivity which result, for example, from the selection rules for momentum conservation). The high energy cutoff of the EDC typically indicates the valence band edge, E_V, the position of which can be determined to within 0.1 eV with reference to the Fermi level, E_F, of a metal in contact with the sample. Figure 3.1a indicates that knowledge of E_F–E_V is sufficient to determine the barrier height and extent of bandbending. As the metallic layer thickens, the bandbending may be monitored by following the shift of the core levels, that is, the electrons emitted from the low-lying atomic-like energy levels. The incident photon energy in these studies is typically in the soft X-ray region, from a synchrotron source, so as to maintain a short escape depth for electron emission with acceptable sensitivity. It is important to differentiate between core level shifts due to bandbending and chemical shifts due to an altered chemical environment. The relative change in amplitudes of the column III and column V core level signals also indicates the extent of dissolution and surface segregation of these species. The width of the core level peaks is employed to determine the onset of metallicity, which may occur with submonolayer coverage due to metal atom clustering. This measurement is of central importance in studies of

MIGS behavior. The criteria for determining the onset of metallization has been debated in the recent literature.[14] In the photoemission experiment, stray light,[15] and even the probing photon source,[16] can induce a photovoltage which alters the bandbending. The latter effect invalidated a number of experiments which had measured the effect of doping on bandbending at low temperatures. At room temperature, the effect is generally negligible owing to the relatively large generation-recombination currents in the surface depletion region.

Figure 3.3 illustrates the application of XPS to a study of the NiAl/AlAs/GaAs system.[17] The NiAl/AlAs heterostructure is one of the stable and epitaxial metallic layer systems to attract recent interest.[18] In this case, the thin interfacial AlAs layer suppresses the Al-Ga exchange reaction between NiAl and GaAs. The Ga and As $3d$ core-level positions are monitored with respect to the Fermi level, as determined by rechecking a gold reference. The clean spectra at the top of the figure show a doublet structure due to spin-orbit splitting. Subsequent spectra show contributions from shifted doublets due to interfacial reaction. The Schottky barrier height is determined by the substrate (near-surface) core-level peak shifts. It is necessary to have a sufficiently large escape depth (~2 nm in this case) and sufficient energy resolution so as to properly separate the substrate and interfacial components. The figure indicates that annealing temperatures near 500 °C are required to obtain a high (0.9 eV) barrier height.

A related technique which probes the energy distribution of occupied and unoccupied surface states is surface photovoltage spectroscopy (SPS), also known as the Kelvin probe method. In this technique, a Au probe is placed in close proximity to the semiconductor surface. The resultant capacitor is illuminated by sub-bandgap light which populates or depopulates the surface states. The contact potential difference is monitored as a function of photon energy by mechanically vibrating the Au probe and measuring the capacitively induced current. Since the surface state positions are determined with respect to the valence or conduction band edges, the technique is insensitive to the anomalous photovoltaic effect that can hamper photoemission experiments.[19]

Optical emission by electron bombardment (cathodoluminescence or CL) or laser irradiation (photoluminescence or PL) may be employed to monitor, in situ, the evolution of interfacial and bulk defect levels. Vitturo et al.[11] report such a study for GaAs in which the electron energy is varied to alter the probing depth, and difference spectra are employed to aid in the separation of bulk and interfacial contributions. The As capping and decapping procedure is also examined in this study. The related technique of inverse photoemission spectroscopy (IPS), in which the impinging electrons have energies on the order of 10 eV rather than 1 keV, has also been employed to monitor the evolution of empty surface and interface states.[20] Electron energy loss spectroscopy (EELS), which measures the energy loss of electrons reflected from the surface, is finding increasing application. Stevens et al.[21] employ EELS to monitor the surface excitation (SE) peak due to excitation from the Ga $3d$ core level to the empty Ga dangling bond. Persistence of this peak

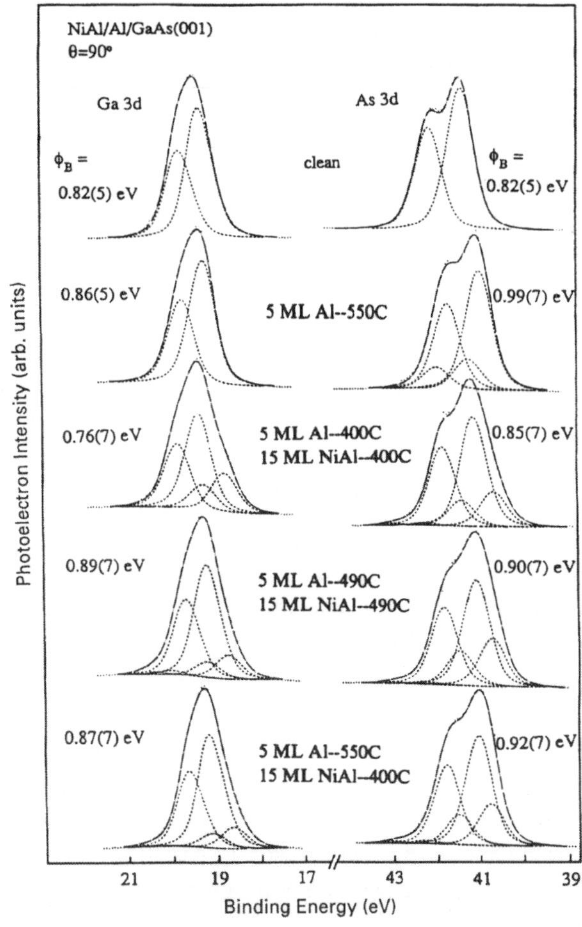

Figure 3.3 Ga and As 3*d* X-ray photoelectron spectra obtained at normal emission for GaAs(001), Al/GaAs(001), and NiAl/Al/GaAs(001) under different growth conditions. The Schottky barrier heights (shown in eV) were determined from core-level spectra. (From S. A. Chambers.[17])

after several monolayers of metal deposition is a clear indication of metal clustering. In this study, LEED I–V (intensity-voltage) spot profiles serve as indicators of subtle changes in the surface unit cells. In the EELS study of Cs/GaAs(110) by DiNardo et al.,[22] the onset of metallization is indicated by features of the surface plasmon loss and the development of the metallic continuum. A striking result in this study is the insulating nature of the interface at coverages up to one monolayer. In a related scanning tunneling microscopy (STM) study by First et al.,[23] linear chains formed by two adjoining rows of Cs atoms are observed on the GaAs(110) surface. STM is also employed to obtain spectroscopic information on filled and empty electronic states by varying the sample bias voltage.[24] The energy resolution is typically superior to that obtained by IPS. The related technique of ballistic electron emission microscopy (BEEM) is discussed below.

Subsequent Metal Deposition

The aforementioned techniques are generally employed in situ to monitor deposition in the submonolayer to several monolayer regime. Thicker depositions are typically examined ex situ. Fine control of multilayer deposition requires monitoring temperature, ambient pressure, and deposition rate. The latter quantity is often measured with a quartz oscillator microbalance,[25] which senses the mass of the evaporated film on the quartz crystal. The mechanical stress in the metal film, which can adversely affect the underlying semiconductor, may be assessed, for example, by deposition onto SiO_2 microcantilever beams, noting the deflection at the end of the beam.[26]

3.3 Unpatterned Test Structures

Electrical Characterization

The periphery of a metallic contact may exert some influence on its electrical properties (see Section 3.4), but the planar interior region generally has the dominant influence. Large-area or wholly unpatterned metallized regions facilitate characterization simply by increasing the volume for analysis. Large-area test patterns for I–V and C–V measurements in which the effect of the periphery is suppressed are described in the next section. The discussion here is confined to characterization techniques in which no patterning is employed.

Prior to metallization, the conductivity of an epitaxial semiconductor layer may be measured by the four point probe method,[1] in which current passed through the outer two probes induces a voltage at the inner two probes proportional to the sheet resistivity. The technique can of course be applied to the deposited metal film as well. The effect of capping or passivation on the starting surface can be assessed by the contactless laser-pumped carrier lifetime bridge, as described by Yablonovitch et al.[27] Light-induced conductivity at the surface is probed by rf induction. Lifetime, band-bending, and surface recombination velocity are accessible in this manner. The technique may also be applicable for very thin metallizations.

BEEM is an extension of the STM technique which can map microscopic variations of the Schottky barrier height with near-atomic resolution.[28] A voltage is applied to the gate metal while tunneling current injected from the STM tip into the semiconductor substrate is monitored. Momentum conservation dictates that only electrons striking the metal semiconductor interface at near-normal incidence can be collected, thereby permitting high spatial resolution. By the same token, topographical gradients associated with hillocks at the interface can strongly affect the BEEM signal. Metal gate thicknesses on the order of 10 nm allow ballistic (i.e., no scattering) transmission of electrons. Spectroscopic information can be obtained by, for example, sweeping the gate voltage while holding the tunneling (tip) current constant. Spectra are normally obtained under modest vacuum conditions, and for metals other than gold, UHV sample transfer or in situ surface cleaning may be required to maintain a clean metal surface. A BEEM investigation of the Au/GaAs interface has elucidated the effects of surface preparation and deposition conditions on interfacial homogeneity.[29]

Concentration Profiling

Depth profiles of elemental concentrations in metallic contacts are routinely obtained by techniques such as Rutherford backscattering spectrometry (RBS), AES, and secondary ion mass spectrometry (SIMS). The utility of RBS to determine stoichiometry, interdiffusion, and surface segregation is illustrated in the review of Palmstrøm and Morgan.[2] Following RBS analysis, AES sputter profiling, for example, can be performed to obtain high-resolution depth profiles.[30]

The depth resolution obtained by sputter profiling is limited by ion bombardment-induced surface roughening, ion knock-on mixing, preferential sputtering, chemical segregation, overlayer nonuniformity, and (in SIMS) non-steady-state behavior at the interface. To obtain high resolution within the semiconductor, it is highly advantageous to sputter from the backside, rather than through the metal layer, provided the semiconductor can be uniformly etched to a manageable thickness.[31–33] For this purpose, epitaxial structures have been prepared containing an etch-stop layer and periodic marker layers for calibration and alignment of subsequent depth profiles. For GaAs substrates, an etch stop layer of $Al_xGa_{1-x}As$ ($x > 0.4$) is resistant to an $NH_4OH/H_2O_2/H_2O$ etch and can subsequently be removed with an HF etch. For InP, an InGaAs etchstop layer is resistant to the substrate HCl etch, but can be removed in $H_2SO_4/H_2O_2/H_2O$. This dual etching technique results in a near atomically smooth surface. The PdGe contact described in Figure 3.2 has been examined by backside SIMS, as illustrated in Figure 3.4. The position of the original interface is determined within 2 nm with the aid of the $Al_{0.1}Ga_{0.9}As$ marker layers. The Ge concentration is observed to drop below a concentration of 10^{19} cm^{-3} within 15 nm of the interface, placing severe constraints on the doping model for ohmic conduction in this contact. Strong SIMS matrix effects cause the Ge signal to be much enhanced upon reaction with Pd, distorting the concentration profile but providing information on the extent of alloy formation.

Electron Microscopy

Both plan-view and cross-sectional transmission electron microscopy (TEM) have proven invaluable in understanding metal-compound semiconductor reactions on a 1 nm scale. The ability to attain real-space images at atomic or lattice resolution, microdiffraction patterns, and elemental information (e.g., energy dispersive X-ray analysis), all on the same (10 nm)^3 volume of material, provides insight into interfacial reactions that is not available by other techniques. In combination with a high-sensitivity depth profiling technique such as those discussed above, TEM data can elucidate the detailed reaction mechanisms that alter the electronic properties

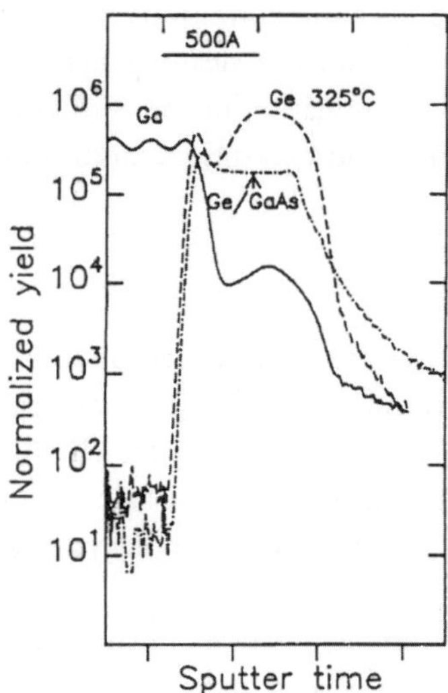

Figure 3.4   Backside SIMS spectra from Ge/Pd/GaAs contact. Spectra from two samples are shown with scales adjusted so that Ga profiles from the two samples overlap. This scaling was made possible by using a GaAs/Al$_{0.1}$Ga$_{0.9}$As superlattice as the substrate material (note oscillations in Ga profile). The profile labeled "Ge/GaAs" is the Ge signal from an unannealed amorphous Ge layer deposited on the marker-layer substrate. This profile is to be compared with the Ge signal (labeled "Ge 325C") from a Ge/Pd contact on the same substrate after annealing at 325 °C for 30 min. Spectra were obtained with a 2.3 keV O$_2{}^+$ beam. The depth scale is for GaAs only. (After Marshall et al.[6])

of interfaces. In Figure 3.5, we again employ the PdGe ohmic contact example[6] to illustrate the role of TEM in contact studies. The samples imaged are identical to those examined in the SIMS study of Figure 3.4. With the aid of the (Al,Ga)As marker layers, Figure 3.5a shows that the upper GaAs layer, after exposure to air and subsequent capping with amorphous Ge, is 9.3 ± 0.3 nm thick. Starting with the uncapped marker layer substrate, a Ge/Pd metallization was deposited and then annealed at 200 °C for 30 min. The resulting reaction to form Pd$_4$GaAs consumes a uniform layer of GaAs of thickness 3.3 ± 0.5 nm. At this stage, the contact displayed rectifying behavior. Subsequent annealing at higher temperatures was found to induce the decomposition of this interfacial phase concomitant with the formation of a low-resistance ohmic contact. As shown in Figure 3.5c, all but about 1 nm of the original GaAs layer had regrown epitaxially. As discussed above, the SIMS data show that this regrown layer contains a high concentration of Ge. The SIMS data also accounts for the "missing" 1 nm of GaAs which was found to have diffused into the metallization. Thus, by combining backside SIMS, electrical measurements, and TEM, fundamental mechanisms of ohmic contact formation can be revealed.

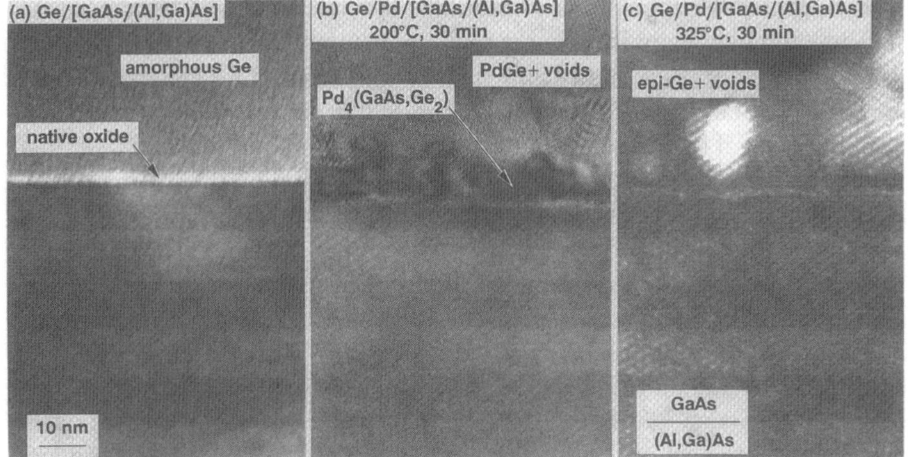

Figure 3.5 Cross-sectional TEM images of the interfacial regions at different stages of the Ge/Pd/GaAs reaction sequence: (*a*) an amorphous Ge layer is deposited on the marker-layer substrate to indicate the original position of the chemically prepared substrate surface; (*b*) after the initial reaction of Pd with GaAs surface at 200 °C to form $Pd_4(GaAs,Ge_2)$; (*c*) after the reaction at 325 °C—the initial $Pd_4(GaAs,Ge_2)$ layer formed between room temperature and about 200 °C has decomposed, and a thin layer of GaAs has regrown on the substrate. The "missing" 1 nm of GaAs has diffused into the contact, as determined by SIMS studies calibrated by comparison with PdGe films implanted with Ga^+ and As^+ ions. (After Marshall et al.[6])

3.4 Patterned Test Structures

Barrier Height

The Schottky barrier height (Figure 3.1*a*) is the most important characteristic of rectifying contacts used as controlling gates in field-effect devices. In general, a high barrier height is desirable, especially in logic circuits, where the tolerable variation in driver threshold is increased when the barrier height of the gate contacts is increased. In the monolayer coverage regime discussed in Section 3.2, the barrier height may be probed by various in situ techniques. Contact metallizations in devices, however, are generally too thick for such measurements. Barrier heights at buried interfaces can be thoroughly characterized using a combination of I–V and C–V measurements and internal photoemission (IPE) spectroscopy (these characterization techniques are reviewed in Reference 34). Each technique gives a value of the effective barrier height based on a simple model of the band diagram and transport characteristics of the Schottky barrier. In principle, all three measurements on an ideal Schottky barrier should yield the same barrier height to within a few tens of meV. However, deviations from ideal behavior affect each measurement differently. All three measurements combined will usually provide sufficient information to identify the source of the nonideality.

The simplest of the measurements is IPE, which relies on the fact that electrons excited in a metal film by the absorption of a photon can be collected by the semiconductor only if the photon energy is greater than the barrier height. Fowler[35] arrived at the approximate relation $Y = C(h\nu-q\phi_b)^2$, where Y is photocurrent per absorbed photon, h is incident photon energy, $q\phi_b$ is barrier height, and C is a constant.

To perform the measurement, a Schottky barrier is illuminated either from the backside of the wafer or from the metallized side, assuming the metal film is thin (≤ 20 nm). Contacts to the metallized side and the back of the substrate permit the measurement of photocurrent as a function of photon energy. One then plots $\sqrt{Y}$ as a function of photon energy (Figure 3.6a). Ideally, this plot should be linear in the photon energy range between the barrier height (more precisely, $q\phi_b + 3kT$, where k is the Boltzmann constant and T is temperature) and the semiconductor bandgap. This linear portion of the curve is extrapolated to the abscissa to yield the barrier height. This method does not depend on contact area and gives roughly an areal average value of the barrier height. With some deconvolution, IPE spectra can also be useful for evaluating more complicated structures with more than one barrier in series.

The most commonly employed measurement technique is the forward-biased I–V measurement. Assuming charge is transported by thermionic emission over the barrier, the current density J is related to the applied voltage V by $J(V) = A^*T^2\exp\{(q/kT)(V - \phi_n)\}$, where A^* is the Richardson constant. Plotting $\log(J)$ versus V should yield a straight line with a slope of 1.0 (Figure 3.6b). Deviations from ideal thermionic emission behavior can result from recombination in the space charge region, edge leakage, nonuniform barrier heights, thin interfacial tunneling barriers, excessive series resistance, and other factors. The excessive series resistance appears as a decreasing slope in $\log(J)$ versus V at high applied voltages as J becomes linear in V. Most other deviations from ideal behavior are manifested as a region of slope less than 1 at low or moderate applied voltages. These deviations can be described by an empirical fitting factor known as the ideality factor, n (V/n is substituted for V in the above equation). In practice, high quality diodes show $n < 1.1$ over more than six orders of magnitude in current. Imperfections that result in small regions with low effective barrier height can dominate the I–V characteristics. Diodes shunted by edge leakage, locally low barrier height regions, or recombination generally yield n approaching a value of 2. Edge leakage can often be minimized by mesa etching. Other contributions to high values of n are generally representative of the transport behavior of the contact within the pattern periphery.

Despite the complications involved in interpreting nonideal I–V data, this technique is the most widely used method for Schottky barrier characterization. This preference is directly attributable to the fact that the I–V measurement is most representative of actual device operation.

The C–V technique provides a third option for barrier height evaluation. In this measurement, the capacitance of a Schottky barrier is measured as a function of the

　　　　　　　　　　　　　　　　　　　　CONTACTS　　Chapter 3

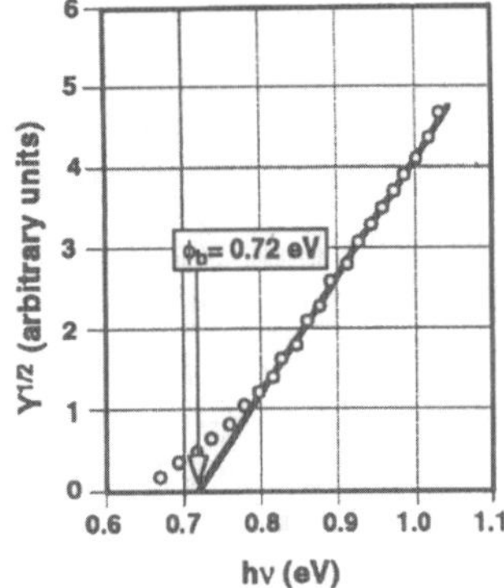

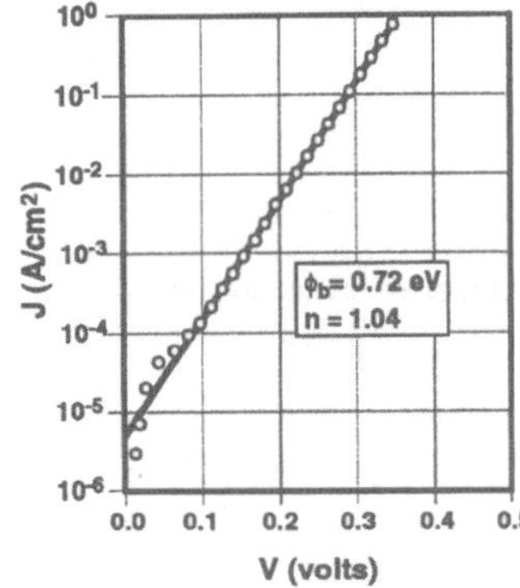

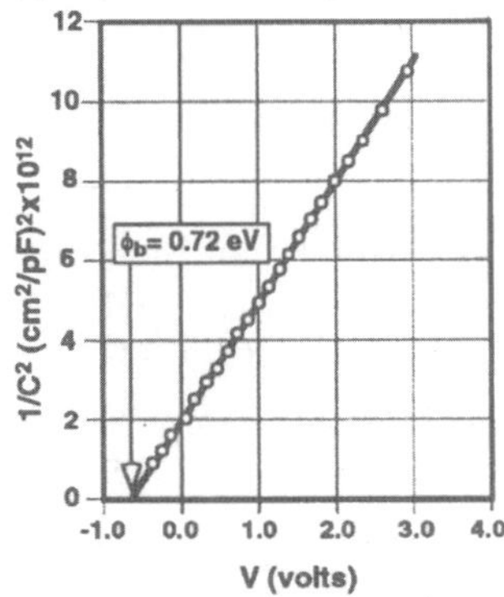

Figure 3.6 Schematic diagrams illustrating the determination of Schottky barrier height using the three techniques described in the text: (*a*) IPE spectrum plotted as $\sqrt{Y}$ versus photon energy to yield an estimate for barrier height. Y is the photocurrent per absorbed photon; (*b*) forward I–V characteristic, illustrating the derivation of the barrier height and ideality factor based on the thermionic emission model; and (*c*) plot of C^{-2} versus applied voltage. The barrier height is determined by the intercepts of the straight lines in all three plots. Data shown are hypothetical.

reversed biased voltage. Both the barrier height and the net substrate doping level can be extracted from a plot of C^{-2} versus V (Figure 3.6*c*). Unlike the I–V technique, the C–V method gives an average value of the barrier height. The C–V method, however, is very sensitive to interfacial layers and doping variations near the interface. Consequently, C–V data often provides complementary information to I–V and IPE measurements, especially for nonideal diodes.

Contact Resistance

The most important characteristic of an ohmic contact is the contact resistance. As for the barrier height, several methods are available for evaluating contact resistance, including the transfer length model (also known as the transmission line method or TLM) and the Cox and Strack method.[35] The TLM measurement is the most widely implemented, since it is most compatible with typical device structures and can be included as part of a device or circuit mask set. In any of the contact resistance measurements, the goal is to obtain a meaningful value of contact resistance by factoring out other contributions to the measured resistance. For example, in the Cox and Strack method, an array of circular dots of different diameters is patterned so that the contact resistance can be separated from the contributions of the spreading resistance in the semiconductor and the series resistance associated with the substrate and the probes. In the TLM, a line of contact pads separated by a range of gap lengths is patterned onto a conducting transmission line fabricated by etching a conducting epilayer grown on a semi-insulating substrate. Measurements of the resistance between pads as a function of gap length allows one to factor out the epilayer sheet resistance. If the metallurgical reactions do not alter the sheet resistance under the contact pads, a "specific contact resistance," given in ohms·square centimeter, can be extracted.

Morphology

Contact morphology is becoming an increasingly important issue, as devices are becoming smaller and more complex. Ideally, a Schottky or ohmic contact should be uniform on the nanometer scale, should consume a minimum of substrate material during annealing, and should be dimensionally stable throughout processing. In the previous section, we described the application of cross-sectional TEM combined with marker-layer substrates to evaluate the interaction depth, uniformity, and morphology of large-area contact metallizations. By patterning the metallization, edge definition and lateral spreading during annealing can be evaluated with TEM or SEM. Such studies are becoming increasingly important as the lateral dimensions of contacts decrease. For example, recent work by Shih et al.[36] has shown that the conventional Au-Ge/Ni ohmic contact metallization to n-GaAs spreads laterally by as much as 0.5 µm during contact annealing. The spreading is irregular, leading to variations in source resistance of field effect transistors as well as limiting the degree to which such devices may be scaled down in size. This observation and similar measurements of contact spiking have motivated much of the recent efforts to develop shallow solid-phase contact metallizations.

References

1   E. H. Rhoderick and R. H. Williams. *Metal-Semiconductor Contacts*, 2nd ed. Clarendon Press, Oxford, 1988.

2   C. J. Palmstrøm and D. V. Morgan. In *Gallium Arsenide*. (M. J. Howes and D. V. Morgan, Eds.) Wiley, New York, 1985, chapt. 6.

3   W. Mönch. *Rep. Prog. Phys.* **53**, 221, 1990.

4   T. Sands. *Mat. Sci. Eng.* **B1**, 289, 1989.

5   M. Murakami. *Mat. Sci. Rep.* **5**, 273, 1990.

6   E. D. Marshall, S. S. Lau, C. J. Palmstrøm, T. Sands, C. L. Schwartz, S. A. Schwarz, J. P. Harbison, and L. T. Florez. *Mater. Res. Soc. Symp. Proc.* **148**, 163, 1989; E. D. Marshall. Ph.D. thesis, University of California, San Diego, 1989.

7   C. J. Palmstrøm, T. L. Cheeks, H. L. Gilchrist, J. G. Zhu, C. B. Carter, and R. E. Nahory. *Mat. Res. Soc. Symp. Proc.* **EA-21**, 63, 1990.

8   S. K. Donner, K. P. Caffey, and N. Winograd. *J. Vac. Sci. Tech.* **B7**, 742, 1989.

9   H. H. Farrell and C. J. Palmstrøm. *J. Vac. Sci. Tech.* **B8**, 903, 1990.

10   D. E. Aspnes, W. E. Quinn, and S. Gregory. *J. Vac. Sci. Tech.* **A9**, 870, 1991.

11   R. E. Viturro, J. L. Shaw, L. J. Brillson, J. M. Woodall, P. D. Kirchner, G. D. Pettit, and S. L. Wright. *J. Vac. Sci. Tech.* **B6**, 1397, 1988.

12   E. Yablonovitch, C. J. Sandroff, R. Bhat, and T. Gmitter. *Appl. Phys. Lett.* **51**, 33, 1987; C. J. Sandroff, R. N. Nottenburg, J.-C. Bischoff, and R. Bhat. *Appl. Phys. Lett.* **51**, 33, 1987.

13   S. D. Offsey, J. M. Woodall, A. C. Warren, P. D. Kirchner, T. I. Chappell, and G. D. Pettit. *Appl. Phys. Lett.* **51**, 441, 1987.

14   W. E. Spicer and R. Cao. *Phys. Rev. Lett.* **62**, 605, 1989, and responses.

15   E. Yablonovitch, B. J. Skromme, R. Bhat, J. P. Harbison, and T. J. Gmitter. *Appl. Phys. Lett.* **54**, 555, 1989.

16   M. H. Hecht. *J. Vac. Sci. Tech.* **B8**, 1018, 1990.

17   S. A. Chambers. *J. Vac. Sci. Tech.* **B7**, 737, 1989.

18   T. Sands, C. J. Palmstrøm, J. P. Harbison, V. G. Keramidas, N. Tabatabaie, T. L. Cheeks, R. Ramesh, and Y. Silberberg. *Mat. Sci. Rep.* **5**, 99, 1990.

19   L. Burstein, J. Bregman, and Y. Shapira. *J. Appl. Phys.* **69**, 2312, 1991.

20   R. Ludeke, D. Straub, F. J. Himpsel, and G. Landgren. *J. Vac. Sci. Tech.* **A4**, 874, 1986.

21   K. Stevens, L. Soonckindt, and A. Kahn. *J. Vac. Sci. Tech.* **A8**, 2068, 1990.

22   N. J. DiNardo, T. M. Wong, and E. Plummer. *Phys. Rev. Lett.* **65**, 2177, 1990.

23   P. N. First, R. A. Dragoset, J. A. Stroscio, R. J. Celotta, and R. M. Feenstra. *J. Vac. Sci. Tech.* **A7**, 2868, 1989.

24   C. K. Shih, R. M. Feenstra, and P. Martensson. *J. Vac. Sci. Tech.* **A8**, 3379, 1990.

25   C. Benndorf, G. Keller, H. Seidel, and F. Thieme. *J. Vac. Sci. Tech.* **14**, 819, 1977.

26   W. A. Strifler and C. W. Bates. *J. Mater. Res.* **6**, 548, 1991.

27   E. Yablonovitch and T. Gmitter. *IEEE Trans. Electron Device Lett.* **EDL-6**, 597, 1985.

28   W. D. Kaiser and L. D. Bell. *Phys. Rev. Lett.* **60**, 1406, 1988.

29   L. D. Bell, M. H. Hecht, W. J. Kaiser, and L. C. Davis. *Phys. Rev. Lett.* **64**, 2679, 1990.

30   E. Relling and A. P. Botha. *Appl. Surf. Sci.* **35**, 380, 1989.

31   I. R. Hill, W. M. Lau, G. R. Yang, and R. A. North. *Surf. Int. Anal.* **11**, 596, 1988.

32   C. J. Palmstrøm, S. A. Schwarz, E. D. Marshall, E. Yablonovitch, J. P. Harbison, C. L. Schwartz, L. Florez, T. J. Gmitter, L. C. Wang, and S. S. Lau. *J. Appl. Phys.* **67**, 334, 1990.

33   S. A. Schwarz, C. J. Palmstrøm, C. L. Schwartz, T. Sands, L. G. Shantharama, J. P. Harbison, L. T. Florez, E. D. Marshall, C. C. Han, S. S. Lau, L. H. Allen, and J. W. Mayer. *J. Vac. Sci. Tech.* **A8**, 2079, 1990.

34   G. Y. Robinson. In *Physics and Chemistry of III–V Compound Semiconductor Interfaces.* (C. W. Wilmsen, Ed.) Plenum, New York, 1985, chapt. 2.

35   R. H. Fowler. *Phys. Rev.* **38**, 45, 1931.

36   Y.-C. Shih, M. Murakami, E. L. Wilkie, and A. C. Callegari. *J. Appl. Phys.* **62**, 582, 1987.

4

Dielectric Insulating Layers

H. H. WIEDER

Contents

4.1 Introduction

The dielectric properties of insulating layers on III–V compound semiconductors depend, in part, on the inherent bulk characteristics and, to a much greater extent, the physical and chemical interfacial reactions between these layers and their semiconductor surfaces.[1] Insulating layers can be made by electrochemical anodization, thermal or plasma oxidation, or by inert ion bombardment of the semiconductor surfaces. Alternatively, synthetic layers can be deposited upon the semiconductor surfaces to reproduce, as closely as possible, the characteristic properties of the silicon/silicon oxide system, such as the capacitance–voltage (C–V) measurements shown in Figure 4.1. No comparable results have been obtained as yet on III–V compound semiconductors with fundamental bandgaps larger than 1 eV.

Better technological procedures than those presently available are required in order to reduce the density of surface states and interfacial traps, and to decrease the surface and interfacial recombination velocities of III–V compound semiconductors. In general, the composition, structure, and interfacial stoichiometry of such insulator-semiconductor structures have been characterized in greater detail

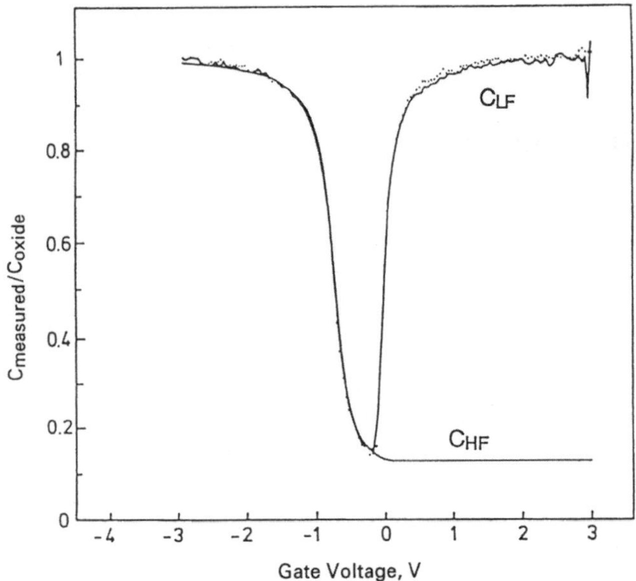

Figure 4.1 **Experimentally measured high frequency (1 MHz) and low frequency (20 Hz), capacitance versus voltage curves compared to theoretically calculated equilibrium C–V values for *p*-type Si/SiO$_2$ MOS structure.**

than have their electrical properties, much of which is incomplete. For example, relatively few C–V measurements have been made under equilibrium or near-equilibrium (quasi-static) conditions. As a rule, 1 MHz C–V measurements are considered to represent true "high frequency" measurements (i.e., no surface states respond at or above this frequency). Analysis of the high frequency C–V data does not, in itself, reveal the dynamic properties of surface or interface states, and the conductance method, which is applied with much success to the silicon/silicon oxide system, is only rarely used for the characterization of III–V compound metal–insulator–semiconductor (MIS) structures.

Nevertheless, a great deal of progress has been made during the past decade, specifically on the two technologically important semiconductors GaAs and InP. These are described, in detail, in the following sections.

4.2 Oxides and Oxidation

Native oxides of the III–V compound semiconductors are, as a rule, compositionally inhomogeneous: they contain various amounts of the atomic constituents of the semiconductor as well as impurities introduced during their synthesis. Their crystalline phase, lattice order, and morphology are dependent on the parameters of their growth procedures. In order to overcome the inadequacy of native or thermally grown oxides for device applications, synthetic dielectric layers such as SiO_2 or Si_3N_4 were deposited on III–V semiconductors. In most instances, an interfacial, thin (~20 Å thickness) native oxide is present between such insulators and the semiconductor surfaces. Schwartz[2] has reviewed various ternary phase diagrams of group III and group V elements and oxygen, and described the manner in which such diagrams can be used to analyze oxide-substrate reactions on III–V semiconductors. He also arrived at the conclusion that thermal oxidation of the arsenides and antimonides of the III–V compounds will always lead to low resistivity structures as a consequence of the thermodynamic stability of elemental arsenic and antimony at their interfaces. For phosphorus-based compounds, such as GaP and InP, purely thermodynamic considerations suggest that single-phase orthophosphates should form on thermal oxidation in air or in O_2. Such layers are polycrystalline and also have low resistivities. Anodic and plasma-grown oxide layers of arsenides and antimonides are also subject to oxide-substrate reactions which release elemental arsenic or antimony.

Homomorphic dielectric layers on GaAs have been made by microwave plasma oxidation and rf gaseous oxidation, as well as by anodic oxidation in an aqueous solution of glycol and other electrolytes. The anodic oxide consists principally of Ga_2O_3 and As_2O_3. The free oxide surface is As-deficient, while the oxide in the vicinity of the semiconductor interface contains a gradient of excess As which may occupy up to 20% of the total oxide thickness.

Enhancement of surface oxides on GaAs formed by ultraviolet and ozone excitation was investigated by Lau et al.[3] by means of X-ray photoelectron spectroscopy[4] (XPS). Their data confirm that these oxides consist of mixed-phase Ga and As oxides. Desorption of Ga oxide phases occurs between 550 and 600 °C, and desorption of As oxide phases and oxygen transfer from As to Ga occurs between 250 and 500 °C. Other methods used in the attempt to grow device-quality native oxides at low temperatures include the exposure of GaAs surfaces simultaneously to O_2 and electron beams,[5] the use of more reactive oxidizers,[6, 7] and the exposure of GaAs surfaces to plasma excitation[8] of O_2. Although such techniques can increase substantially the rate of formation of the first few oxide monolayers, they are, with the exception of plasma oxidation, ineffective in promoting the growth of thick layers—the resulting oxides are usually inhomogeneous. The growth of ~50 nm thick oxide layers on (100)- and (110)-oriented GaAs using a high kinetic energy atomic oxygen beam was undertaken by Hoffbauer et al.[9] Such layers were found, by means of XPS, to be uniform in composition and fully oxidized, and Raman spectroscopy indicated no appreciable lattice disorder at the semiconductor interface.

An alternative to oxidation is nitridation of GaAs surfaces. The large binding energy of N_2 molecules prevents their dissociation on GaAs surfaces. Gourier et al.[10] and Friedel and co-workers[11–13] investigated the nitridation of GaAs surfaces by the induction of a nitrogen plasma. Atomic nitrogen, such as produced by electron impact, may also promote nitridation.[14] The interaction of atomic nitrogen with (110)-oriented GaAs surfaces—investigated by Berger et al.[15] using Auger electron spectroscopy[16] (AES), low-energy electron-loss spectroscopy[17] (LEELS), and ultraviolet photoemission spectroscopy[18] (UPS)—indicates chemisorption of nitrogen at low exposures, while for larger exposures a GaN surface layer is formed by way of an anion exchange reaction. In contrast with these data are the results obtained by Soukiassian et al.,[19] who found that alkali metals, such as in a rubidium-covered GaAs surface, can increase substantially the room temperature nitridation by producing mixed nitrides, such as $GaAs_{1-x}N$, GaN, and GaN_{1+x}, and much higher Ga nitridation states. A necessary condition seems to be the reactivity of the alkali metal-semiconductor interface suggesting the mediation of surface defects in the semiconductor which are produced by the rubidium overlayers.

Anodic layers have been grown on InP using both aqueous and nonaqueous electrolytes. The electrical properties of such layers are very sensitive to water vapor. Radio frequency plasma oxidation of InP has been made using an rf-generated oxygen plasma. Auger spectroscopic investigations show the bulk of this oxide to be P deficient; it tends to pile up in the vicinity of the interface. The properties of thermally grown InP oxides in dry oxygen were investigated by Wager and Wilmsen.[20] They found that the oxides composed of 70–75% In_2O_3 and 25–30% P_2O_5 grew very slowly at temperatures below 340 °C and rapidly above this temperature. Thermal oxidation of InP begins with the formation of a layer of $InPO_4$, which is the expected oxide for equilibrium growth. Indeed, for temperatures greater than 650 °C the entire oxide layer is $InPO_4$, while oxides formed in lower temperatures contain, in addition, an admixture of In_2O_3. Thermally grown oxides have a low resistivity, of the order of 10^8 and 10^9 Ω·cm, and a static dielectric constant of 10.3. The resistivity of such layers can be increased[21] to between 10^{11} and 10^{12} Ω·cm by performing the oxidation in P_2O_5 vapor at a pressure of 0.01 to 0.2 atm. Liliental et al.[22] found that such oxides, grown at 350 °C, consist of 3–7 nm thick $InPO_4$ and a 6 nm thick outer layer of In_2O_3. Thermal and photon-assisted reactions of ammonia, silane, and oxygen with InP were investigated by Houzay et al.[23] They found that photo-assisted nitridation of InP with NH_3 is feasible, and that photo-assisted and thermally induced reactions with SiH_4 remove, in part, the native oxide-forming covalent bonds between the Si atoms and the substrate.

4.3 Heteromorphic Insulators

Heteromorphic or synthetic insulators such as SiO_2 or Si_3N_4 have been deposited on III–V compound semiconductors by means of procedures such as pyrolysis or sputtering.[1] Such layers are usually amorphous and more nearly homogeneous than homomorphic layers. However, the high temperatures used for their synthesis tend to alter the stoichiometry of the semiconductor surfaces on which they are deposited.

There is considerable interest in the development of low-temperature procedures for the synthesis of heteromorphic insulators. One of these depends on the creation of short-lived chemically reactive species produced by means of a photo-enhanced or plasma-enhanced chemical vapor deposition (PECVD). Lucovsky et al.[24] found that low-temperature PECVD deposition of SiO_2 produces a thin film of native oxide at the semiconductor interface which can play a significant role in determining the electrical properties of the interface. In the case of GaAs, this oxide is predominantly Ga_2O_3. Such subcutaneous oxidation appears[25] as well in the PECVD deposition of SiO_2 on InP. In this case, XPS reveals the presence of an $InPO_4$ layer between the deposited SiO_2 and InP. Photolytic deposition of silicon nitride on GaAs using 193 nm excimer radiation acting on SiH_4, NH_3, and He within a reaction chamber was investigated by Eisele et al.[26] They found that the lowest substrate temperature for obtaining near-stoichiometry was 250°C; the growth rate of the layers was 600/800 Å/min for a pulse rate of 50 Hz and energy of 70–90 mJ.

Nissim et al.[27] provide a comparative evaluation of the low-temperature, low-pressure, ultraviolet photo-assisted (UVCVD) synthesis of insulating layers on semiconductors and the high-temperature, transient, "flash" pyrolytic deposition of such layers. There seems to be some merit for considering both techniques for the deposition of insulating layers on both InP and $In_{0.53}Ga_{0.47}As$.

Comparatively few investigations have been made thus far on the synthesis of anodic and SiO_2 deposited heteromorphic layers on ternary alloy semiconductors, such as $In_xGa_{1-x}AsyP_{1-y}$. Chen et al.[28] found that its anodic oxides consist of non-uniformly distributed regions which contain Ga_2O_3 and In_2O_3; Ga_2O_3, In_2O_3, and As_2O_5; Ga_2O_3, In_2O_3, As_2O_5, and P_2O_5; In_2O_3 and P_2O_5; and a native oxide transition region present under the SiO_2 layer.

4.4 Chemical Modification of GaAs Surfaces

Oxygen-induced bandbending on *n*-doped GaAs by a fractional monolayer of oxygen is sufficient to pin the Fermi level 0.65 eV below the conduction band minimum (CBM); but on *p*-type GaAs, it takes considerably more oxygen to pin the Fermi level 0.55 eV above the valence band edge (VBM). Equilibrium Fermi level pinning is attributed to a high density of fast surface states, N_{ss}, present at all of the interfaces[29] with homomorphic and heteromorphic insulators deposited on GaAs.

Quasi-static C–V measurements as well as high-frequency C–V measurements made on GaAs MIS structures indicate that the total change in surface potential is $\Delta V_s = 0.4$ V and that the Fermi level can be displaced from ~0.6 eV below the CBM to ~1.0 eV below the CBM; therefore, neither accumulation or inversion is feasible due to an $N_{ss} > 10^{13}$ cm^{-2}/eV, which pins the Fermi level near midgap, as shown in Figure 4.2.

Fermi level pinning positions determined on MIS structures made on (100)-oriented GaAs[30] are essentially the same as those of (110)-oriented ultra-high-vacuum-cleaved surfaces coated with metals such as Cs or coated with a fractional monolayer of oxygen. This suggests that it is the density of the surface states on the semiconductor rather than the properties of the insulator which determine, at least to first order, Fermi level pinning.

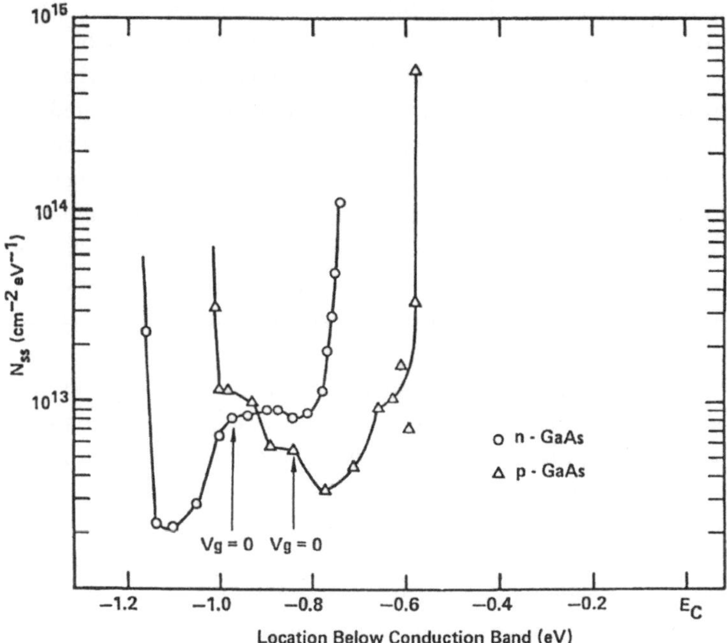

Figure 4.2 Surface state density as a function of energy within the band-gap of *n*-type and *p*-type GaAs derived from high frequency measurements. (After L. G. Meiners, "Dielectric-Semiconductor Interfaces of GaAs and InP," Colorado State University, SF-19, 1979.)

The chemical modification of GaAs surfaces coupled with the removal of the native oxides and residual contaminants was assumed to lead to unpinning of the Fermi level pinning, or at least to a reduction in N_{ss}. However, the XPS data of Münch[31] and Ismail et al.[32] suggest that not only surface states but also traps within the oxide may affect the equilibrium Fermi level position.

Offsey et al.[33] used a photochemical method to obtain a substantial reduction of the nonradiative surface recombination velocity of GaAs. It is based on the dissolution of the fraction of native oxide containing As_2O_3 and of any As clusters present at the oxide-semiconductor interface. A reduction in the density of surface or interface states might accompany the formation of a $GaAs/Ga_2O_3$ heterojunction. However, C–V measurements[34] made on MIS structures with SiO_2 superposed on Ga_2O_3 layers are not conclusive.

A surface modification introduced by Yablonovitch and co-workers[35, 36] consists of depositing, from an aqueous Na_2S solution, a polycrystalline layer of $Na_2S \cdot H_2O$ on GaAs. Immediately after deposition, the surface recombination velocity, normally of the order of 10^7 cm/s, is reduced by several orders of magnitude; however, after 20 to 30 min this effect disappears. A fast, 60-fold increase, followed by a time-dependent decrease, of the current gain of heterojunction bipolar transistors was obtained by Sandroff et al.[37] using this technique. The claim that it also causes a reduction in N_{ss} to ~5 × 10^{11} cm^{-2} was challenged by Besser and Helms,[38] who found, using a surface conductivity measurement technique, that the Fermi level remains pinned. A similar conclusion was reached by Hasegawa et al.[39] and by Spindt et al.[40] using XPS. It appears that such sulfur treatment moves the Fermi level towards the VBM from ~0.75 eV to 0.5 eV above the VBM. Measurements using deep level transient spectroscopy (DLTS) also provide evidence that, while states near the middle of the bandgap are removed, those within 0.3 eV of the VBM are not affected by such treatment.[41]

Carpenter et al.[42] found similar reductions in the recombination current of GaAs homojunction diodes using an ammonium sulfide solution. This type of treatment also degrades with time but appears to be more durable than that using $Na_2S \cdot H_2O$. Cowans et al.[43] suggest that this process produces a slight Ga enrichment and leaves ~0.6 monolayer of the sulfide to inhibit oxidation. The XPS investigations of Sandroff et al.[44] indicate that the passivating layers of As_xS_y formed by the reaction of either of the two aqueous sulfides decompose in the presence of light, leaving on the surface a residue of As_2O_3 on the $Na_2S \cdot H_2O$-covered surface, while $(NH_4)_2S$-covered surfaces form a two-phase layer of As_2O_3 and As_2S_3, the latter being more photochemically and oxidation stable.

Further details of the chemistry and the electrical properties of the As_2S_3-GaAs amorphous-crystalline heterojunction were provided by Yablonovitch et al.,[45] who found that the GaAs surface recombination velocity is reduced to ~15 000 cm/s. The resistivity of the amorphous layer is ~10^{18} $\Omega \cdot$cm, and the interfacial charge, which is negative, is estimated to be ~6 × 10^{11} cm^{-2}.

A different form of surface chemical passivation was reported by Turco et al.,[46] who used a solution of Na_2Se in NH_4OH to produce a polycrystalline layer of Na_2Se on GaAs. Such Se-coated surfaces were found to have a 10- to 20-fold larger photoluminescence efficiency than the same uncoated specimens, as shown in Figure 4.3. This was attributed to a reduction in the density of the surface recombination centers due, probably, to a gallium selenate surface layer. Sandroff et al.[47] found that by first depositing elemental selenium from an aqueous solution on a GaAs surface followed by its immersion in a sodium sulfide solution it is possible to modify the recombination characteristics of GaAs surfaces, at least temporarily, so that their photoluminescence efficiency[48] increases by a factor of 400 compared with untreated surfaces. Of equal interest is the temporal, thermal, and environmental stability of such chemically passivated surfaces. Chambers and Sundaram[49] used a metal-organic chemical vapor deposition (MOCVD) process to produce selenium passivated GaAs surfaces by soaking MBE-grown GaAs layers in hydrogen-diluted H_2Se. From their angle-resolved XPS measurements, they infer a substantial reduction in bandbending, which they attribute to the formation of a heterojunction between GaAs and a thin pseudomorphic $GaAs_xSe_{1-x}$ overlayer; the reduction of interface states is assumed to occur as a consequence of the elimination of dangling bonds at the heterojunction interface.

4.5 Indium Phosphide–Insulator Interfaces

InP–insulator interfaces are in most respects different from those of GaAs primarily because N_{ss} at midgap is smaller by more than one order of magnitude and because a variety of semiconductor surface preparation and chemical modification techniques are available for eliminating Fermi level pinning. Low-coverage (110)-oriented oxygen-exposed n-type InP surfaces are depleted due to an acceptor level located at 0.45 eV relative to the CBM. Increasing the oxide thickness moves the equilibrium Fermi level to ~0.2 eV relative to the CBM. Removal of this oxide returns the surface Fermi level to 0.45 eV. This suggests that at least some of the shallow donor levels might be located in the oxide rather than on the InP surface. Two different types of interface states are present at InP–insulator interfaces: (1) slow states located at or within the insulator, which produce a C–V hysteresis and a slow drift of the MIS capacitance with time and (2) fast interface states, usually in the range $N_{ss} = 10^{11}$ to 10^{13} cm^{-2}/eV, which have a U-shaped surface potential dependence. The origin of these states is attributed by Hasegawa and Ohno[50] to interfacial disorder-induced gap states, and by others[51] to donor-like phosphorous vacancies and acceptor-like cation antisite defects. C–V measurements made on (100)-oriented InP MIS structures indicate that the equilibrium surface Fermi level of untreated surfaces is consistent with those measurements made on (110)-cleaved bulk n- and p-type (110)-cleaved InP.[52]

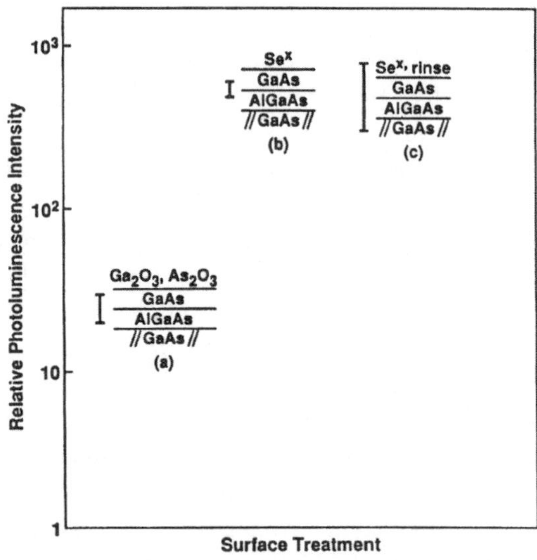

Figure 4.3 Photoluminescence response at room temperature from an undoped GaAs layer using a 50 mW 588 nm Ar-ion laser with (*a*) native oxide, (*b*) oxide terminated surface after Se treatment, and (*c*) Se treatment after a 10 min rinse in 18 MΩ water. (After Reference 46.)

Chemical treatment before deposition or growth of a homomorphic or heteromorphic insulating layer on InP can have significant advantages, as shown in Figure 4.4. Wilmsen et al.[53] have investigated, by means of XPS and AES, the sulfurization of InP surfaces. They found that S replaces surface phosphorus or that it fills surface vacancies forming an In_2S_3/InP heterojunction. Heteromorphic MIS structures made by means of such a process by Iyer et al.[54] have N_{ss} values in the range between 10^{10} and 10^{11} cm^{-2}/eV. Low-pressure thermally induced sulfurization of InP surfaces was described by Rakib et al.,[55] who found that layers with the composition $InP_{0.82}S_{4.0}$ can be grown at temperatures of 270 to 280°C. Such layers were found to have N_{ss} values between 2×10^{11} and 5×10^{11} cm^{-2}/eV near midgap and typical breakdown fields of 7×10^6 V/cm.

Chemical modification of InP surfaces prior to the deposition of Al_2O_3 insulating layers, by subjecting them to an As overpressure of 10^{-4} to 10^{-6} torr or by etching in a H_3AsO_4 solution, was investigated by Chave et al.[56] They found that such treatment reduces or eliminates the frequency dispersion of the capacitance in accumulation and reduces or eliminates the hysteresis observed on 1 MHz C–V curves. Arsenic stabilization of InP surfaces prior to the deposition of Al_2O_3 insulating layers was also described by Gendry et al.[57] They confirmed the reduction in C–V hysteresis, and suggested a decrease in the density of interfacial traps. They estimated N_{ss} to be between 10^{11} and 5×10^{11} cm^{-2}/eV at the N_{ss} (Vs) minimum located ~0.3 eV below the CBM.

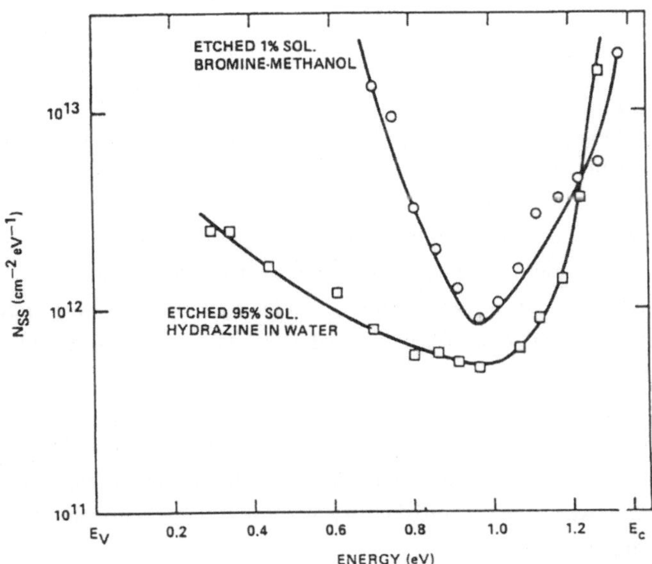

Figure 4.4 Density of surface states as a function of energy determined from high-frequency capacitance measurements on InP/SiO_2 MIS structures with different surface preparations prior to the deposition of the insulator. (After Reference 25.)

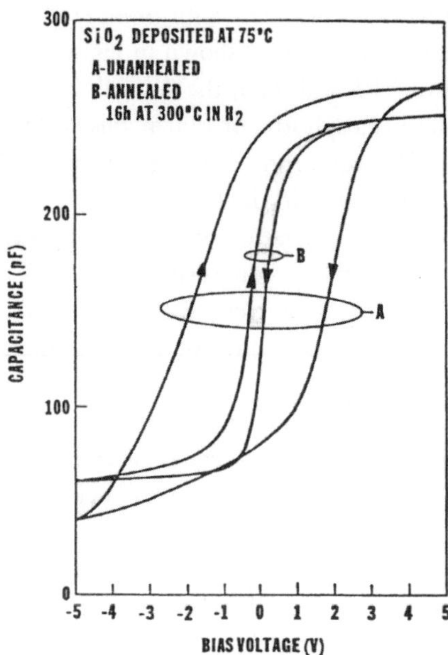

Figure 4.5 High-frequency (5 MHz) C–V hysteresis of an InP/SiO$_2$ MIS capacitor before and after annealing in hydrogen at 300 °C for 16 h. (After Reference 61.)

The changes produced in the chemical composition and the equilibrium Fermi level of n- and p-type InP surfaces by their reaction with sodium sulfide and with ammonium sulfide solutions have been characterized by Lau et al.[58] using XPS. They found that in sulfide solutions, all phosphorus sulfide and most of the indium sulfide dissolved, leaving about a monolayer of indium sulfide on the InP surfaces. Such treatment appears to reduce the density of donor states in the upper half of the bandgap but might introduce a compensating density of acceptor states. If air oxidation products of sulfides are present on p-type InP, then heating in vacuum to 300 °C introduces strong surface inversion. Deposition of a silicon nitride layer on such chemically modified surfaces produces an N_{ss} of the order of 10^{11} cm^{-2}/eV but leaves a high density of slow states.

Annealing InP in an overpressure of phosphorus[59, 60] prior to the deposition of a synthetic insulator, produces intermediate polyphosphide layers such as In(PO$_3$)$_x$. These have a large bandgap and a high resistivity and are chemically stable; their interfacial properies are, however, not as satisfactory as those of heteromorphic layers grown at low temperatures.

Gardner and Narayan[61] have described the synthesis of SiO$_2$ layers at temperatures between 75 °C and 170 °C using a photo-stimulated disproportionation process of SiH$_4$ and N$_2$O to deposit SiO$_2$ layers on InP, followed by long-term annealing of the layer in hydrogen at 3000 °C. Such a low reaction temperature is

     DIELECTRIC INSULATING LAYERS   Chapter 4

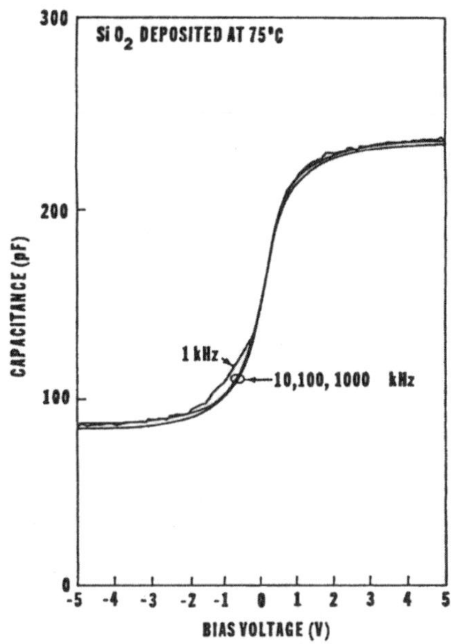

Figure 4.6 Frequency dispersion of the capacitance, after annealing, of the same specimen described in Figure 4.5. (After Reference 61.)

made possible by resonant energy absorption from an external radiation source acting on mercury vapor present in the same reactor. This provides the requisite energy for disproportionation of the source materials. The effects of annealing in hydrogen and the frequency dispersion of the C–V curves are shown in Figures 4.5 and 4.6, respectively.

A different method for a low-temperature deposition of silicon dioxide on InP at a pressure of 500 to 650 mtorr and with substrate temperatures in the range of 120 to 150 °C was described by Iyer et al.[62] They used an indirect PECVD process with an rf power input to the plasma source of 14 W. Under these conditions the deposition rate was ~5 Å/min, and the insulating layers had refractive indices and resistivities comparable to those grown at higher temperatures. In contrast with these techniques which require external energy sources for reducing reaction temperatures, Vaccaro et al.[63] used a low-pressure, commercially available reactor modified so that the gases are mixed before their entrance into the reaction chamber; the gases then flow over and under a hot platen, thus extending the residence time of the reactive species. Based on their data, Vaccaro et al. arrived at the conclusion that the SiO_2 deposition process is reaction limited at low temperatures and diffusion limited at high temperatures. Controlling the oxygen partial pressure and allowing the source gases to preheat appear to be critical factors for optimizing this process. Such layers, grown on InP at or below 100 °C, exhibited C–V hysteresis of less than 0.1 V and resistivities of the order of 10^{15} Ω·cm. Remote plasma-enhanced chemical vapor deposition (RPECVD) in a reaction chamber containing nitrogen, oxygen, and tetraoxysilane, with P vapor introduced in the initial stages of growth, was also used by Kulisch and Kassing[64] to deposit SiO_2 layers on InP at 200 °C. They found that the hysteresis and temporal drift of their InP MIS structures were associated with deep, spatially and energetically distributed insulator traps within the interfacial native oxide, and that the density of these traps can be reduced by more than one order of magnitude compared to that of MIS structures made by high-temperature deposition procedures if the semiconductor surface stoichiometry is preserved. C–V and DLTS measurements indicated that by using RPECVD and low-temperature procedures it is possible to move the Fermi level over most of the fundamental bandgap, and to reduce the density of slow states from $\sim 4 \times 10^{12}$ cm^{-2}/eV to the order of 10^{11} cm^{-2}/eV.

A model for the slow drift of the flat-band voltage with reference to InP MIS field effect transistors was developed by Goodnick et al.[65] in terms of thermally assisted electron tunneling from the InP surface to a conducting interfacial In_2O_3 layer within the native oxide. Assuming plausible values for the relevant parameters, they were able to simulate the time dependence of the drift; however, their model is unable to explain the absence of such a drift at low temperatures. Van Staa et al.[66] have proposed a different model for this drift in terms of traps distributed in space and in energy within the insulator in the immediate vicinity of the interface. Electrons in these traps communicate with the InP conduction band by tunneling into the insulator, where they are captured or emitted by a thermally assisted tunneling process. Their own DLTS measurements as well as those of Meiners[67] appear to agree with this model. The slow drift of the flat-band voltage and C–V hysteresis is attributed by Van Vechten and Wager[68] to the hopping of one of the four nearest neighboring In atoms into a phosphorus vacancy, thus creating a defect complex consisting of an In vacancy and an InP antisite defect. In general, the use of low growth temperatures, the use of P during the growth of heteromorphic layers, and long-term annealing in hydrogen tend to reduce, though not completely eliminate, such drift.

4.6 Heterojunction Quasi-Insulator Interfaces

A large bandgap, totally depleted semiconductor in contact with a lower bandgap n- or p-doped semiconductor can perform the function of a quasi-insulator in an MIS structure. An isotype or anisotype heterojunction used in this manner requires large band-edge discontinuities so that only the dielectric properties of the quasi-insulator are significant and the conduction current across its interface is negligible. The C–V characteristics of such a heterojunction may be described in terms of an equivalent MIS structure. Fixed interface charges and fast interface states related to interfacial lattice defects can affect the C–V characteristics of such heterojunctions.

Casey et al.[69] and Gosard et al.[70] have measured the I–V and C–V characteristics of such $GaAs/Ga_xAl_{1-x}As$ heterojunctions, and Solomon et al.[71] have described the properities of GaAs/Ga$_{0.6}$Al$_{0.4}$As/GaAs capacitors. They found a negative fixed charge in the GaAlAs of ~4 × 10^{16} cm^{-2} which produces a flat-band voltage shift of 0.16 eV and N_{ss} of the order of 10^{10} cm^{-2}/eV. The role of interfacial defects in such Ga$_x$Al$_{1-x}$As/GaAs heterojunctions was investigated by Zur and McGill.[72] They show that defect densities of ~10^{12} cm^{-2} can affect the position of the surface Fermi level; however, interface state densities of the order of 10^{13} to 10^{14} cm^{-2} are required to pin the Fermi level. The DLTS measurements of As et al.[73] indicate the presence of deep electron traps at the interface between n-isotype Ga$_x$Al$_{1-x}$As/GaAs quantum wells with x in the range between 0.24 and 0.39. Such states are considered to be associated with defects present within the first 10–20 nm of Ga$_x$Al$_{1-x}$As.

Boron nitride indium phosphide heterojunctions were made by Yamaguchi and Minakata[74] by the pyrolysis and reaction of NH$_3$, B$_2$H$_6$, and PH$_3$. They found the boron nitride layers to be B$_x$N, with x in the range between 1.3 and 2. MIS structures exhibited a breakdown strength larger than 3 × 10^6 V/cm, a resistivity of the order of 10^{15} Ω·cm, and N_{ss} ~ 10^{10} cm^{-2}/eV. Paul et al.[75] used a Q-switched ruby laser with an energy density of 2.5 J/cm^2 applied to BN polycrystalline wafers to evaporate and deposit boron nitride layers on InP. MIS structures made in this manner were found to have an anticlockwise C–V hysteresis, and N_{ss} = 6 × 10^{10} cm^{-2}/eV at 0.5 eV below the CBM. However, such layers are hygroscopic, and N_{ss} appears to increase to 4.1 × 10^{11} cm^{-2}/eV following post-deposition annealing.

Mizuta and co-workers[76, 77] and Fujieda et al.[78] have grown aluminum nitride layers on n-type GaAs and on InP using a low-temperature MOCVD process. Trimethyl gallium and trimethyl aluminum were used as the group III sources and hydrazine as the nitrogen source. Deposition of cubic AlN layers on GaAs was obtained at temperatures above 220 °C and GaN layers about 450 °C. C–V and DLTS measurements made on MIS structures led to identification of two discrete traps: one at $E_c - 0.7$ eV and another at $E_c - 0.9$ eV. Between the conduction band and midgap, N_{ss} = 10^{12} cm^{-2}/eV, in spite of the relatively large lattice mismatch between AlN and GaAs; the Fermi level can be displaced over more than 0.8 eV within the GaAs fundamental bandgap.

The electrical properties of In$_{0.52}$Al$_{0.48}$As/In$_{0.53}$Ga$_{0.47}$As MIS heterojunctions grown on InP substrates by means of MBE were investigated by Hong et al.[79] using frequency-dependent admittance measurements. They found N_{ss} to be in the range between 10^{12} and 8 × 10^{12} cm^{-2}/eV, with emission time constants from these states of 100 and 700 ns. The electrical properties of In$_x$Ga$_{1-x}$As/GaAs MIS heterojunctions were evaluated theoretically by Jeong et al.[80] by solving Poisson's equation with the interface trap density modeled in terms of a box charge located at the hetero-junction interface. Qualitative agreement between theory and experiment was obtained for traps located at $E_c - 0.13$ eV with a concentration of 8 × 10^{10} cm^{-2}/eV. DLTS measurements indicate the location of a trap at $E_c - 0.13$ eV and at $E_c - 0.17$ eV, with respective capture cross sections of 2 × 10^{-14} and 1 × 10^{-15} cm^2.

The electrical properties of In$_x$Al$_{1-x}$As/InP MIS structures, in which the quasi-insulator is deliberately mismatched in order to gain, with decreasing x, a larger

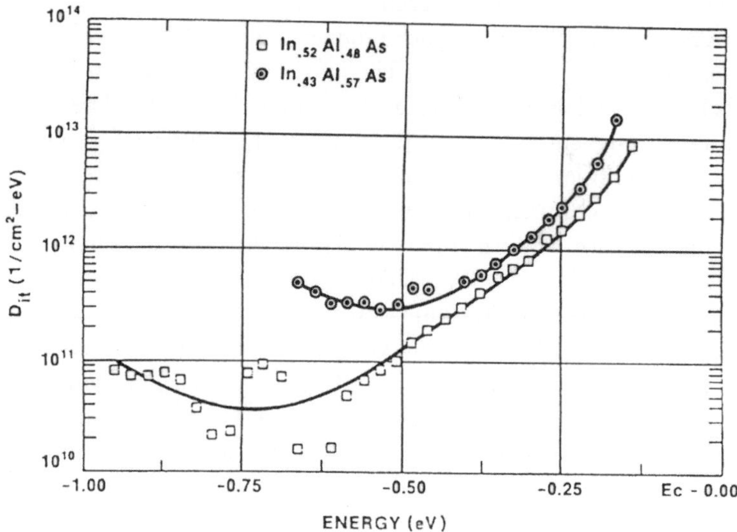

Figure 4.7 Interface state density as a function of energy for the lattice-matched heterojunction InP/In$_{0.52}$Al$_{0.48}$As and for a similar but mismatched structure with an In$_{0.43}$Al$_{0.57}$As quasi-insulator derived from high-frequency C–V measurements. (After Reference 81.)

conduction band-edge discontinuity and hence a lower reverse bias leakage current, were investigated by Hanson and Wieder.[81, 82] The equilibrium surface potential is $E_c - 0.05$eV, and the Fermi level can be displaced by an applied electric field over most of the indium phosphide bandgap, as shown in Figure 4.7, without, however, achieving inversion. No long-term drift of the flat-band potential was found in such structures, suggesting the absence of deep level interfacial traps with long time constants.

A different method of making such quasi-insulators, described by Casey and co-workers[83, 84] is based on the use of oxygen-doped Al$_{0.5}$Ga$_{0.5}$As/GaAs MIS structures. It employs oxygen-doped Al$_{0.5}$Ga$_{0.5}$As as the insulator of GaAs MIS structures with MBE growth conditions so chosen that oxygen is readily incorporated into the Al$_{0.5}$Ga$_{0.5}$As but not in the GaAs. C–V measurements made on such structures, shown in Figures 4.8 and 4.9, exhibited no hysteresis, and DLTS measurements indicated the presence of a relatively low interface trap density. Admittance measurements as a function of frequency and temperature reveal a deep level at 0.64 ± 0.4 eV below the CBM, with a concentration of 4×10^{16} cm^{-3}. For the AlAs fraction $x = 0.5$, the fundamental bandgap is indirect, $E_g = 2.0$ eV, the electron mobility of the n-type material is ~200 cm^2/V·s, and the resistivity of thin layers is high enough for conduction currents to be negligible. Admittance measurements made on such MIS structures suggest an equivalent circuit of the GaAs depletion capacitance in series with the quasi-insulator capacitance in low

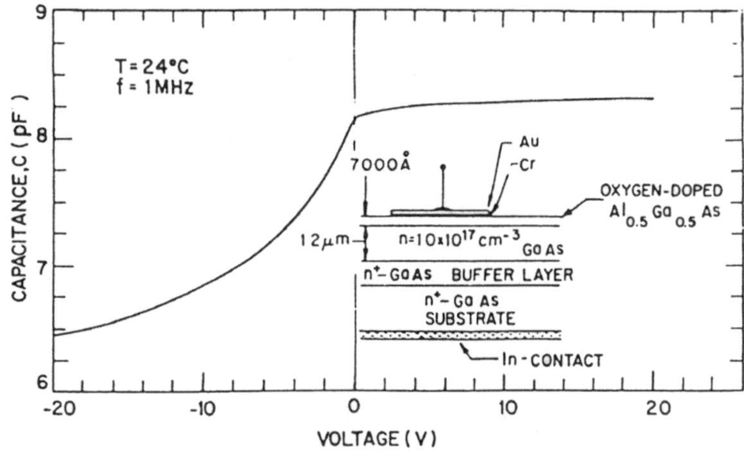

Figure 4.8   High-frequency capacitance as a function of gate voltage for a GaAs MIS structure with an AlGaAs quasi-insulator. (After Reference 69.)

to moderate reverse bias with flat-band expected near zero bias; no inversion is obtained in high fields due to leakage of minority carriers through the quasi-insulator.

Andre et al.[85] have investigated the properties of similar MIS structures, with the $Al_{0.5}Ga_{0.5}As$ doped with either oxygen or chromium. In accumulation, the measured capacitance corresponds to that of the calculated quasi-insulator thickness and dielectric constant. The Cr-doped specimens were found to have lower leakage currents and higher breakdown fields—greater than 7×10^5 V/cm—than the oxygen-doped layers.

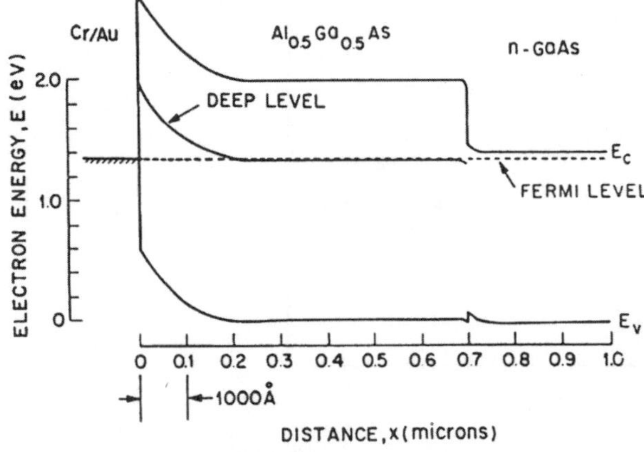

Figure 4.9   Energy band diagram of MIS heterostructure shown in inset of Figure 4.8. (After Reference 69.)

Pruniaux et al.[86] and Fleming[87] have made quasi-insulating layers on GaAs surfaces by bombarding them with protons of 25 eV energy and 10^{14} cm^{-2} fluence. They found that the lattice disorder produced by the ion bombardment makes such layers essentially semi-insulating, with resistivities of the order of 10^8 Ω·cm and a fixed interfacial charge of ~5 × 10^{11} cm^{-2}. However, C–V measurements made on MIS structures with such layers by Meiners,[88] using frequencies up to 150 MHz, indicated that, contrary to the results of Pruniaux et al., flat-band is not reached and the Fermi level is pinned near midgap. This suggests that Ar-ion bombardment can generate a high enough density of defect-related interface states to pin the Fermi level.

4.7 Epitaxial Fluoride Insulators

A different type of heterojunction, in which one side is effectively an insulator, is that between III–V compounds and the crystalline alkaline earth fluorides, such as CaF_2 and SrF_2. These fluorides sublimate instead of dissociating when evaporated in ultra-high vacuum, and can be deposited, therefore, by MBE as stoichiometric layers. However, the GaAs/SrF$_2$ lattice mismatch is 2.56%. The lattice constants of CaF_2 and SrF_2 bracket those of GaAs. Assuming the validity of Vegard's law, $Ca_{0.44}Sr_{0.56}F_{0.2}$ can provide an adequate lattice match, and Siskos et al.[89] have grown, by MBE, such a solid solution fluoride on (100)-oriented GaAs substrates. Electron diffraction[90] and Raman scattering spectroscopy[91] confirm the epitaxial character and the transitivity of GaAs/(CaSr)$_2$F$_2$ heterojunctions. However, the difference in chemical bonding and the more nearly covalent character and more strongly ionic bonding of the fluorides, as well as the much greater elastic coefficient stiffness of the fluorides, might be responsible for the incoherence of the

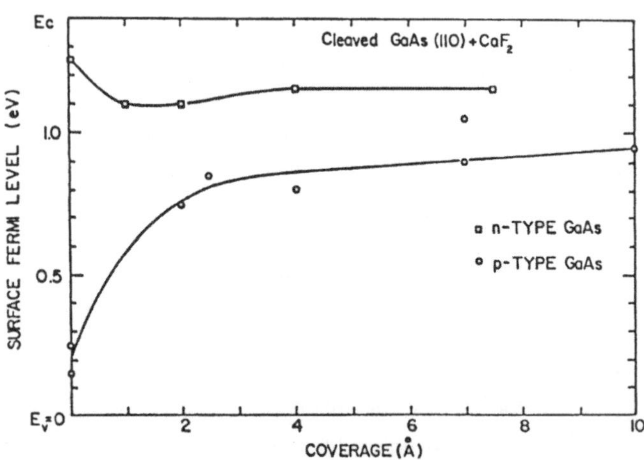

Figure 4.10   Position at equilibrium of surface Fermi levels, a function of coverage, in angstroms, of the cleaved (110)-oriented GaAs surface covered by CaF$_2$. (After Reference 97.)

                    DIELECTRIC INSULATING LAYERS   Chapter 4

interfaces observed with reflected high energy electron diffraction (RHEED) and transmission electron microscopy (TEM).[92–94]

C–V measurements made on CaF_2/GaAs MIS structures by Sinhroy et al.[95] indicate the presence of a large N_{ss} and Fermi level pinning of the GaAs surface. Barriere et al.[96] subjected (100)-oriented GaAs to 1 bar of fluorine at temperatures greater than 300 °C to induce the growth of GaF_3 insulating layers. C–V measurements made on such MIS structures show a small hysteresis at 1 MHz, and suggest the presence of fixed negative interface charges.

The interfaces between (100)-oriented cleaved GaAs surfaces and either CaF_2 or SrF_2 layers deposited upon them in high vacuum were investigated by Mao et al.,[97] who found an initially large bandbending on p-type GaAs and a small bandbending on n-type GaAs, as shown in Figure 4.10. The final Fermi level pinning positions after deposition of about one monolayer of CaF_2 is 0.95 eV above the VBM for p-type GaAs and 0.25 eV below the CBM for n-type GaAs. The band offsets are estimated to be 7.7 eV and 8.0 eV for CaF_2 and SrF_2, respectively. Waho et al.[98] and Waho and Saeki[99] investigated the pseudomorphic growth of CaF_2 and $(CaSr)_2F_2$ on (111) B-oriented GaAs. They found that temperatures higher than 580 °C are required for the formation of Ca-As bonds at the interface. They arrived at the conclusion, based on C–V and isothermal transient capacitance spectroscopic measurements, that the Fermi level is not pinned, and that it can be displaced toward the conduction band to within $E_c - 0.1$ eV, while MIS structures made on (100)-oriented surfaces have their Fermi level pinned. However, their C–V data, shown in Figure 4.11, requires further confirmation; their low frequency C–V data, obtained under illumination at 77 K, cannot be considered to represent equilibrium inversion or accumulation. Such conclusions may also be drawn from the limited C–V data on GaAs/CaF_2 MIS structures measured by Bousetta and Truscott.[100] Based on a capacitance-versus-time drift, they suggest that mobile positive ions initially concentrated in the CaF_2 and electron traps localized in the vicinity of the GaAs interface are responsible for a substantial low-frequency hysteresis.[101]

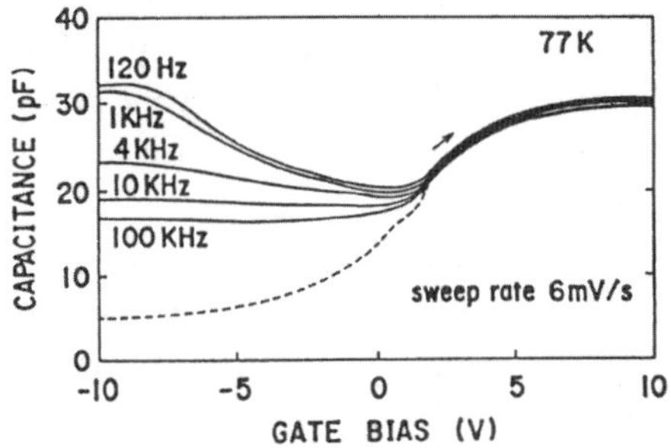

Figure 4.11 Frequency dispersion of the capacitance as a function of voltage of a CaSrF$_2$/GaAs MIS structure under illumination (*solid line*) and in the dark (*dashed line*); hysteresis is not shown. (After Reference 98.)

Fluoridation of InP surfaces has been investigated by Barriere et al.[102] using various reaction temperatures and time durations. The bulk composition of the insulating layers, determined by means of XPS, infrared absorption and Rutherford back-scattering[103] is InF_3. At the fluoride-InP interface, both indium and phosphorus are bound to fluorine. They found that reaction temperatures below 200 °C make the InF_3 layers hygroscopic.

The 5.7% lattice mismatch between BaF_2 and InP and the 6.9% mismatch between CaF_2 and InP can lead to plastic deformation of such layers, with the formation of interfacial lattice defects. Farrow et al.[104] have grown BaF_2 on InP, and Sullivan et al.[105] found that CaF_2 layers can be grown on InP substrates heated to ~350 °C. Double heterostructures of CaF_2/InP/CaF_2, as well as $Ba_xSr_{1-x}F$/InP MIS structures, were made by Tu et al.[106] They found that while the deposition of InP on CaF_2 is accommodated by pseudomorphic growth, the converse does not work. Such double heterostructures were found to have poor electrical properties; their C–V data suggests the presence of deep level traps located either at the heterostructure interfaces or within the insulator. The investigations of Weiss et al.[107] on the CaF_2/InP interface using XPS low-energy electron diffraction[108] and ultraviolet photoelectron spectroscopy[108] indicate that bandbending at the interface results from p-type defects in the CaF_2 with strong Fermi level pinning. The synthesis and some of the properties of $Ba_{1-x}Sr_xF_2$ layers deposited on InP were also investigated by Paul and Bose[109, 110] for x = 0, 0.5, and 1.0. They found that such layers had resistivities of the order of 10^{12} Ω·cm and a breakdown field > 5 × 10^5 V/cm, and that the pretreatment of the InP surfaces in HF prior to deposition of the insulator is beneficial; N_{ss} derived from a Terman-type analysis is of the order of 10^{10} cm^{-2} at its minimum, E_c – 0.47 eV.

4.8 Commentary

Some of the methods intended to provide adequate insulation layers on III–V compound semiconductor surfaces have been concerned with preservation of the interface between the native oxide, or a selected phase of this oxide, and the semi-conductor surface. Others have introduced a controlled modification of the surface chemistry in order to prevent, or to eliminate, Fermi level pinning, and to reduce N_{ss} while avoiding the formation of deep level interfacial traps. Preservation of the native oxide is not an adequate method for eliminating Fermi level pinning in GaAs.

Chemical modification of GaAs surfaces by the chemisorption of S, Se, or Si prior to the deposition of heteromorphic insulating layers can overcome some of the impediments associated with the high surface recombination velocity of this compound. The technique used by Hattangady et al.[111] and by Fountain et al.[112] consists of the deposition of a 10–20 Å Si layer on GaAs prior to a low-temperature, plasma-assisted deposition of a SiO_2 layer. Although the high-frequency C–V hysteresis of such structures suggests the presence of deep levels, their quasi-static C–V data indicates that accumulation as well as inversion has been attained, with a

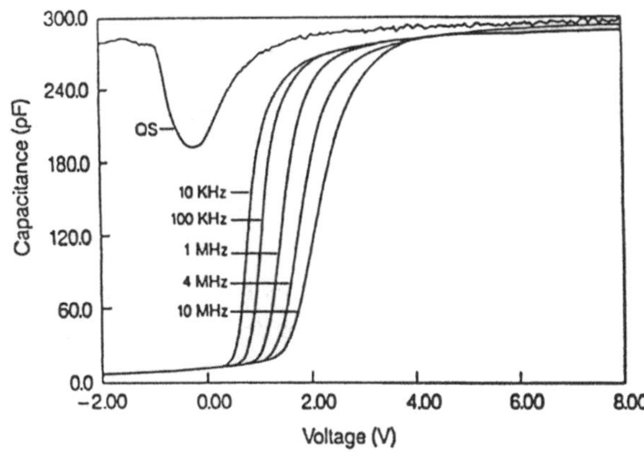

Figure 4.12 **High-frequency dispersion of the capacitance and quasi-static C–V measurements made on a SiO$_2$/SiGaAs MIS structure; hysteresis is not shown. (After Reference 112.)**

surface potential excursion of 0.83 V. They suggested that the reason for such favorable results is that the Si layer provides a heterojunction interface to GaAs, while the superposed SiO$_2$ layer provides the good insulating characteristics desired of an MIS capacitor. Their structure is, therefore, a hybrid which may, in fact, have eliminated Fermi level pinning. Similar results have been obtained by Tiwari et al.[113] who employed MBE to make GaAs/Si MIS heterostructures. In view of the spectroscopic ellipsometric data[114] of Freeouf et al.[115] —who indicate that Si is not present in such structures in an unoxidized state, at least not in quantities necessary to form such a heterojunction—this assumption needs further support. Subsequent work by Freeouf et al.[116, 117] indicates that while at least 2 Å of unoxidized Si is present between the SiO$_2$ and the GaAs, the bandbending of about 0.8 eV and the apparent improvement in the C–V data of such structures has less to do with a reduction in N_{ss} than with a shift in position of these states closer to the CBM.

One may conclude that there is, as yet, no unambiguous evidence of inversion under equilibrium conditions to be obtained on GaAs MIS structures with homomorphic or heteromorphic insulating layers, although in (n^+/p^-) heterojunction structures inversion is clearly obtainable by modulation doping.

The presence or absence of a subcutaneous oxide layer between a heteromorphic insulator and the GaAs surface is not the reason for Fermi level pinning. Neither nitridation nor fluoridation of GaAs surfaces seems to have solved the problem of Fermi level pinning.

While removal of the native oxide affects only slightly the electrical properties of MIS structures, perturbations of the surface stoichiometry of the semiconductor affects them adversely.

The Fermi level cannot be unpinned by providing only an interfacial lattice constant match; if it could, then the large bandgap, large band-edge discontinuity fluorides, lattice-matched to GaAs, should produce an unpinned Fermi level, or at least, a substantial reduction in the N_{ss} of GaAs. Any potential reduction in the surface recombination velocity of GaAs by means of fluoride layers has not been determined as yet.

Fermi level pinning in the vicinity of the CBM, such as occurs in InP and in $In_{0.53}Ga_{0.47}As$, can be removed by chemical pretreatment of the semiconductor surfaces prior to the deposition, at low substrate temperatures, of heteromorphic insulating layers.

All insulator/III–V compound structures investigated thus far, with the exception of lattice-matched heterojunctions, have interface state densities larger by at least one order of magnitude compared with those of the Si/SiO_2 semiconductor/insulator system.

References

1   "Special Issues on Semiconducting III–V Compound MIS Structures." *Thin Solid Films*. Vol. 56, 1979, and Vol. 103, 1983. (C. W. Wilmsen and H. H. Wieder, Eds.) Elsevier, New York. A compendium of the early literature.

2   G. P. Schwartz. *Thin Solid Films*. **103**, 3, 1983.

3   W. M. Lau, R. N. Sodhi, S. Jin, and S. Ingrey. *J. Vac. Sci. Technol* **A8**, 1899, 1990.

4   *Encyclopedia of Materials Characterization*. (C. R. Brundle, C. A. Evans, Jr., and S. Wilson, Eds.) Butterworth-Heinemann, Boston, 1992, p. 282.

5   I. Lopez de Ceballos, M. C. Munoz, J. M. Goni, and J. L. Sacedon. *J. Vac. Sci. Technol*. **A4**, 1621, 1986.

6   K. A. Bertness, T. T. Chiang, C. E. McCants, P. H. Mahowald, A. K. Wahi, T. Kendelewicz, I. Lindau, and W. E. Spicer. *Surf. Sci.* **185**, 544, 1987.

7   P. A. Bertrand. *J. Electrochem. Soc.* **132**, 973, 1985.

8   H. Sommer, W. John, and A. Meisel. *Surf. Sci.* **178**, 179, 1986.

9   M. A. Hoffbauer, J. B. Cross, and V. M. Bermudez. *Appl. Phys. Lett.* **57**, 2193, 1990.

10   S. Gourier, L. Smit, P. Friedel, and P. K. Larsen. *J. Appl. Phys.* **54**, 3993, 1983.

11   P. Friedel and S. Gourier. *Appl. Phys. Lett.* **42**, 509, 1983.

12   P. Friedel, J.-P. Landesman, P. Boher, and J. Schneider. *J. Vac. Sci. Technol.* **B5**, 1129, 1987.

13   P. Friedel and J.-P. Landesman. *Phil. Mag.* **B55**, 711, 1987.

14   F. Bozso and P. Avouris. *Phys. Rev.* **B38**, 3937, 1988.

15   A. Berger, D. Troost, and W. Monch. *Vacuum*. **41**, 669, 1990.

16   *Encyclopedia of Materials Characterization*. (C. R. Brundle, C. A. Evans, Jr., and S. Wilson, Eds.) Butterworth-Heinemann, Boston, 1992, p. 310.

17   *Encyclopedia of Materials Characterization.* (C. R. Brundle, C. A. Evans, Jr., and S. Wilson, Eds.) Butterworth-Heinemann, Boston, 1992, p. 324.

18   *Encyclopedia of Materials Characterization.* (C. R. Brundle, C. A. Evans, Jr., and S. Wilson, Eds.) Butterworth-Heinemann, Boston, 1992, p. 300.

19   P. Soukiassian, H. I. Starnberg, T. Kendelewicz, and Z. D. Hurych. *Phys. Rev.* **B42**, 3769, 1990.

20   J. F Wager and C. W. Wilmsen. *J. Appl. Phys.* **53**, 5789, 1982.

21   M. Yamaguchi. *Appl. Phys.* **52**, 4885, 1981.

22   Z. Liliental, O. L. Krivanek, J. E Wager, and S. M. Goodnick. *Appl. Phys. Lett.* **46**, 889, 1985.

23   F Houzay, J. M. Moison, C. Licoppe, and Y. I. Nissim. *Vacuum.* 718, 1990.

24   G. Lucovsky, S. S. Kim, D. V. Tau, G. G. Fountain, and R. J. Markunas. *J. Vac. Sci. Technol.* **B7**, 861, 1989.

25   L. G. Meiners. *J. Vac. Sci. Technol.* **19**, 373, 1981; L. G. Meiners.*J. Vac. Sci. Technol.* **21**, 655, 1982.

26   K. M. Eisele, W. Rothemund, and B. Dischler. *Vacuum.* **41**, 1081, 1990.

27   Y. I. Nissim, J. M. Moison, F. Lebland, C. Licoppe, and M. Bensoussan. *Appl. Surf. Sci.* **46**, 175, 1990.

28   K. Chen, L. Qui, and W. Chen. *Chin. Phys.* **4**, 459, 1984.

29   L. G. Meiners and H. H. Wieder. "Semiconductor Surface Passivation." *Mat. Sci. Reports.* **3**, 139, 1988. A review of earlier works. H. H. Wieder. "Properties and Applications of Narrow Bandgap Compound Semiconductors." *Solid State Electron.* **33**, 3333, 1990. A review of the passivation of InSb and InAs.

30   W. E. Spicer, P. W. Chye, P. R. Skeath, C. Y. Su, and I. Lindau. *J. Vac. Sci. Technol.* **16**, 1422, 1979.

31   W. Munch. *Appl. Surf. Sci.* **22/23**, 705, 1985.

32   A. Ismail, J. M. Palau and J. Lassabatere. *J. Appl. Phys.* **60**, 1730, 1986.

33   S. D. Offsey, J. M. Woodall, A. C. Warren, P. D. Kirchner, T. I. Chapell, and G. D. Petit. *Appl. Phys. Lett.* **48**, 475, 1986.

34   A. Paccagnella, A. Callegari, J. Batey, and D. Lacey. *Appl. Phys. Lett.* 57, 258, 1990.

35   E. Yablonovitch, C. J. Sandroff, R. Bhat, T. J. Gmitter. *Appl. Phys. Lett.* **51**, 439, 1987.

36   E. Yablonovitch, B. J. Skromme, R. Bhat, J. P. Harbison, and T. J. Gmitter. *Appl. Phys. Lett.* **54**, 555, 1989.

37 C. J. Sandroff, R. N. Nottenburg, J. C. Bischoff, and R. Bhat. *Appl. Phys. Lett.* **51**, 439, 1987.

38 R. S. Besser and C. R. Helms. *Appl. Phys. Lett.* **52**, 1707, 1988.

39 H. Hasegawa, H. Ishii, T. Sawada, T. Saitoh, S. Konishi, Y. Liu, and H. Ohno. *J. Vac. Sci. Technol* **B6**, 1184, 1988.

40 C. J. Spindt, R. S. Besser, R. Cao, K. Miyano, C. R. Helms, and W. E. Spicer. *Appl. Phys. Lett.* **54**, 1148, 1989.

41 D. Liu, T. Zhang, R. A. LaRue, J. S. Harris, Jr., and T. W. Sigmon. *Appl. Phys. Lett.* **53**, 1059, 1988.

42 M. S. Carpenter, M. R. Melloch, and T. E. Dungan. *Appl. Phys. Lett.* **53**, 66, 1988.

43 B. A. Cowans, Z. Dardas, W. N. Degass, M. S. Carpenter, and M. R. Melloch. *Appl. Phys. Lett.* **54**, 365, 1989.

44 C. J. Sandroff, M. S. Hedge, and C. C. Chang. *J. Vac. Sci. Technol.* **7**, 841, 1989.

45 E. Yablonovitch, T. J. Gmitter, and B. G. Bagley. *Appl. Phys. Lett.* **57**, 2241, 1990.

46 F. S. Turco, C. J. Sandroff, M. S. Hedge, and M. C. Tamargo. *J. Vac. Sci. Technol* **B8**, 856, 1990.

47 C. J. Sandroff, M. S. Hedge, L. A. Farrow, R. Bhat, J. P. Harbison, and C. C. Chang. J. Appl. Phys. **67**, 586, 1990.

48 *Encyclopedia of Materials Characterization.* (C. R. Brundle, C. A. Evans, Jr. and S. Wilson, Eds.) Butterworth-Heinemann, Boston, 1992, p. 373.

49 S. A. Chambers and V. S. Sundaram. *Appl. Phys. Lett.* **57**, 2342, 1990.

50 H. Hasegawa and H. Ohno. *J. Vac. Sci. Technol* **B4**, 1130, 1986.

51 P. Boher, M. Renaud, J. M. Lopez-Villegas, J. Schneider, and J. P. Chane. *Appl. Surf. Sci.* **30**, 1987.

52 W. E. Spicer, P. W. Chye, C. Y. Su, and I. Lindau. *J. Vac. Sci. Technol.* **16**, 1422, 1979.

53 C. W. Wilmsen, K. M. Geib, J. Shin, R. Iyer, D. L. Lile, and J. J. Pouch. *J. Vac. Sci. Technol.* **B7**, 851, 1989.

54 R. Iyer, R. R. Chang, and D. L. Lile. *Appl. Phys. Lett.* **53**, 134, 1988.

55 S. Rakib, M. Gendry, P. Klopfenstein, R. Saoudi, and J. Durand. *Thin Solid Films.* **187**, 309, 1990.

56 J. Chave, A. Chojaa, C. Santinelli, R. Blanchet, and P. Viktorovitch. *J. Appl. Phys.* **61**, 257, 1987.

57 M. Gendry. R. Blanchet, G. Hollinger, C. Santinelli, R. Skheyta, and P. Viktorovitch. *SPIE.* **1444**, 282, 1989.

58 W. M. Lau, S. Jin, X.-W. Wu, and S. Ingrey. *J. Vac. Sci. Technol* **B8**, 848, 1990.

59 K. P. Pande and D. Gutteerrez. *Appl. Phys. Lett.* **46**, 416, 1985.

60 Y. Rombach, J. Joseph, E. Begignat, B. Commere, G. Hollinger, and P. Viktorovitch. *Appl. Phys. Lett.* **49**, 1281, 1986.

61 P. D. Gardner and S. Y. Narayan. *SPIE.* **1444**, 186, 1989.

62 R. Iyer, Z. Zou, C. W. Wilmsen, and D. L. Lile. *SPIE.* **1444**, 202, 1989.

63 K. Vaccaro, B. R. Bennett, J. P. Lorenzo, A. Davis, and H. G. Lipson. *SPIE.* **1444**, 233, 1989.

64 W. Kulisch and R. Kassing. *J. Vac. Sci. Technol.* **B5**, 523, 1987.

65 S. Goodnick, T. Hwang, and C. W. Wilmsen. *Appl. Phys. Lett.* **44**, 453, 1984.

66 P. Van Staa, H. Rombach, and R. Kassing. *J. Appl. Phys.* **54**, 4014, 1983.

67 L. G. Meiners. *J. Vac. Sci. Technol* **19**, 373, 1981.

68 J. A. Van Vechten and J. F. Wager. *J. Appl. Phys.* **57**, 1956, 1985.

69 H. C. Casey, A. V Cho, D. V. Lang, E. H. Nicollian, and P. W. Foy. *J. Appl. Phys.* **50**, 3484, 1979.

70 A. C. Gosard, W. Brown, C. L. Allyn, and W. Wiegman. *J. Vac. Sci. Technol.* **20**, 694, 1982.

71 P. M. Solomon, T. W. Hickmott, H. Morkoc, and R. Fisher. *Appl. Phys. Lett.* **42**, 82, 1983.

72 A. Zur and T. C. McGill. *J. Vac. Sci. Technol.* **B2**, 440, 1984.

73 D. J. As, P. W. Epperlein, and P. M. Mooney. *J. Appl. Phys.* **64**, 2408, 1988.

74 E. Yamaguchi and M. Minakata. *J. Appl. Phys.* **55**, 3098, 1984.

75 T. K. Paul, P. Bhattacharya, and D. N. Bose. *Appl. Phys. Lett.* **56**, 2648, 1990.

76 M. Mizuta. S. Fujieda. Y. Matsumoto, and T. Kawamura. *Japan. J. Appl. Phys.* **25**, 945, 1986.

77 M. Mizuta, S. Fujieda, T. Jitsukawa, and Y. Matsumoto. *Inst. Phys. Conf. Ser.* **83**, 153, 1986.

78 S. Fujieda, M. Mizuta, and Y. Matsumoto. *Japan. J. Appl. Phys.* **25**, L945, 1986.

79 W.-P. Hong, J.-E. Oh, P. Bhattacharya, and T. E. Tiwald. *IEEE Trans. Electron. Dev.* **35**, 1585, 1988.

80 J. Jeong, T. Schlesinger, and A. Milnes. *IEEE Trans. Electron. Dev.* **34**, 1911, 1987.

81 C. M. Hanson and H. H. Wieder. *J. Vac. Sci, Technol.* **B5**, 971, 1987.

82 C. M. Hanson and H. H. Weider. *Thin Solid Films.* **153**, 497, 1987.

83 H. C. Casey, Jr., A. Y. Cho, and E. H. Nicollian. *Appl. Phys. Lett.* **32**, 679, 1978.

84 H. C. Casey, Jr., A. Y. Cho, D. V. Lang, and E. H. Nicollian. *J. Vac. Sci. Technol.* **15**, 1408, 1978.

85 J. P. Andre, C. Schiller, A. Mittonneau, A. Briere, and J. Y. Aupied. *Inst. Phys. Conf. Ser. No.* **65**, 117, 1982.

86 B. R. Pruniaux, J. C. North, and A. V. Payer. *IEEE Electron Dev. Lett.* **ED-19**, 672, 1972.

87 P. L. Fleming. *IEEE Electron Dev. Lett.* **EDL-3**, 104, 1982.

88 L. G. Meiners. *J. Appl. Phys.* **50**, 1154, 1979.

89 S. Siskos, C. Fontaine, and A. Munoz-Vague. *Appl. Phys. Lett.* **44**, 1146, 1984.

90 *Encyclopedia of Materials Characterization.* (C. R. Brundle, C. A. Evans, Jr., and S. Wilson, Eds.) Butterworth-Heinemann, Boston, 1992, p. 264.

91 *Encyclopedia of Materials Characterization.* (C. R. Brundle, C. A. Evans, Jr., and S. Wilson, Eds.) Butterworth-Heinemann, Boston, 1992, p. 428.

92 J. M. Philips and J. M. Gibson. *Mat. Res. Soc. Proc.* **25**, 381, 1984.

93 C. Fontaine. J. Castagne, E. Bedel, and A. Munoz-Vague. *Appl. Phys. Lett.* **64**, 2076, 1988.

94 T. Waho and F. Yanagawa. *J. Crystal Growth.* **95**, 515, 1989.

95 S. Sinhroy, R. A. Hoffman, R. F. C. Farrow, J. D. Warner, and K. Basin. *Proc. Symp. Dielectric Films on Compound Semiconductors.* Vol. 86, Electrochemical Society, Pennington, NJ, 1985, p. 313.

96 A. S. Barriere, G. Coutourier, H. Guegan. T. Sequelong, A. Thabti, P. Alnot, and J. Chazelas. *Appl. Surf. Sci.* **41/42**, 383, 1989.

97 D. Mao, K. Young, A. Kahn, R. Zanoni, J. McKinley, and G. Margaritondo. *Phys. Rev.* **B39**, 12735, 1989.

98 T. Waho, F. Yanagawa, Y. Yamada-Maruo, and H. Saeki. *Solid-State Electron.* **33**, 252, 1990.

99 T. Waho and H. Saeki. *Japan. J. Appl. Phys.* **30**, 221, 1991.

100 A. Bousetta and W. S. Truscott. *J. Appl. Phys.* **68**, 5709, 1990.

101 S. Sinhroy. *Thin Solid Films.* **187**, 231, 1990. A recent review on fluoride/semiconductor and semiconductor/fluoride/semiconductor structures.

102 A. S. Barriere, G. Couturier, G. Gevers, H. Guegan, V. Tournay, D. Bertault, B. Desbat, A. Tressaud, and P. Alnot. *Surf Sci.* **239**, 135, 1990.

103 *Encyclopedia of Materials Characterization.* (C. R. Brundle, C. A. Evans, Jr., and S. Wilson, Eds.) Butterworth-Heinemann, Boston, 1992, p. 476.

104 R. F. C. Farrow, P. W. Sullivan, G. M. Williams, G. R. Jones, and D. C. Cameron. *J. Vac. Sci. Technol.* **19**, 415, 1981.

105 P. W. Sullivan, R. F. C. Farrow, and G. R. Jones. *J. Crystal Growth.* **60**, 403, 1982.

106 C. W. Tu, S. R. Forrest, and W. D. Johnston. *J. Appl. Phys. Lett.* **43**, 569, 1983.

107 W. Weiss. R. Hornstein, D. Schmeisser, and W. Gopel. *J. Vac. Sci. Technol.* **B8**, 715, 1990.

108 *Encyclopedia of Materials Characterization.* (C. R. Brundle, C. A. Evans, Jr., and S. Wilson) Butterworth-Heinemann, Boston, 1992, p. 252.

109 T. K. Paul and D. N. Bose. *J. Appl. Phys.* **67**, 3744, 1990.

110 T. K. Paul and D. N. Bose. *Ferroelectrics.* **102**, 397, 1990.

111 S. V. Hattangady, G. G. Fountain, D. J. Viktavage, R. A. Rudder, and R. J. Markunas. *J. Electrochem. Soc.* **136**, 2070, 1989.

112 G. G. Fountain, R. A. Rudder, S. V. Hattangady, R. J. Markunas, and D. J. Viktavage. *Mat. Res. Soc. Symp. Proc.* **144**, 537, 1989.

113 S. Tiwari, S. L. Wright and J. Batey. *IEEE Electron Dev. Lett.* **EDL-9**, 488, 1988.

114 *Encyclopedia of Materials Characterization.* (C. R. Brundle, C. A. Evans, Jr., and S. Wilson, Eds.) Butterworth-Heinemann, Boston, 1992, p. 401.

115 J. L. Freeouf, J. A. Silberman, S. L. Wright, S. Tiwari, and J. Batey. *J. Vac. Sci. Technol.* **B7**, 854, 1989.

116 J. L. Freeouf, D. A. Buchanan, S. L. Wright, T. N. Jackson, J. Batey, B. Robinson, A. Callegari. A. Paccagnella, and J. M. Woodall. *J. Vac. Sci. Technol.* **B8**, 860, 1990.

117 J. L. Freeouf, D. A. Buchanan, S. L. Wright, T. N. Jackson, and B. Robinson. *Appl. Phys. Lett.* **57**, 1919, 1990.

Other Compound Semiconductor Films

OWEN K. WU and DAVID R. RHIGER

Contents

5.1 Introduction

In the area of semiconductor microelectronics, the II–VI compound semiconductors offer an alternative to the more familiar III–V and group IV materials. Bandgaps of the II–VI's range from slightly negative in HgTe to 4.0 eV in ZnS. Low-temperature electron mobilities of more than 10^6 cm^2/V·s have been demonstrated. Because most of the compounds are miscible, the bandgap and lattice parameter of alloys can be continuously varied. Also, having a direct energy gap, the II–VI's are good photon emitters, as well as efficient photon absorbers.

To exploit the attractive characteristics of the II–VI compounds, researchers are attempting to further both the material technology and the device technology. Where possible, techniques established for those semiconductors having more advanced technologies, such as Si and GaAs, are being adapted, but some materials issues, such as doping or volatility of the constituents, are manifested in unique ways in the II–VI's.

A Focus on HgCdTe

The II–VI material that has thus far received the greatest development effort and found the largest market is $Hg_{1-x}Cd_xTe$, for infrared (IR) detectors.[1–11] Here,

x denotes the mole fraction CdTe in a pseudobinary alloy with HgTe. (For simplicity, the subscripts in the formula are often omitted.) To illustrate the applications of several analytical techniques to II–VI compounds, we have chosen HgCdTe as the focus of this chapter. Closely related materials include HgZnTe, which is also suitable for IR detection,[12, 13] and CdTe and CdZnTe, which form appropriate substrates for thin-film growth of the IR-sensing materials.[14, 15]

Objective and Scope

Our objective is to show how analytical techniques can be applied to solve problems encountered in the development and fabrication of HgCdTe devices. Descriptions of the individual analytical techniques can be found in the lead volume of this series, *Encyclopedia of Materials Characterization*. Because of space limitations, we cannot address every major issue of HgCdTe technology, nor can we describe a device fabrication process in any detail. Instead, we select specific problems related to interfaces and epitaxial layers, with an emphasis on materials rather than devices, and review solutions available in the literature. The examples are selected for illustrative purposes and do not necessarily represent the preferred approach in every laboratory.

Background

HgCdTe is the material of choice for IR detectors because of its variable bandgap, its high quantum efficiency, its capability for operation above liquid nitrogen temperatures, and its potential for enhanced device performance through superlattice and quantum-well based structures.[11] HgCdTe detectors sensitive to wavelengths of 2 to 12 μm have already been demonstrated in several laboratories, and many are now in use for sensing or imaging applications, both civilian and military. Array size and density have steadily increased to meet stringent system demands in a variety of different applications.[2–4] By way of illustration, a state of the art 128×128 HgCdTe focal plane array is shown in Figure 5.1a.[8] The major problems associated with the development of HgCdTe have been materials and process related; these include control of composition, morphology, native defects, precipitates, doping concentrations, crystalline quality, and surface order/cleanliness. Most of these problems have their origin in the relatively weak chemical binding of Hg, and the vulnerability of the alloy surface to ambients and to chemicals during processing.

Three basic kinds of devices have been used as IR detectors: photoconductor (PC), photovoltaic (PV), and metal–insulator–semiconductor (MIS).[16] For instance, PC detectors have been in production for many years. PV and MIS detectors are in various stages of development and pilot production. We have chosen the PV device (specifically the p-on-n double-layer structure) for discussion here because not only is it the most promising structure for high performance long wavelength detection, but also it presents all of the surface and interface issues to be addressed in this chapter.

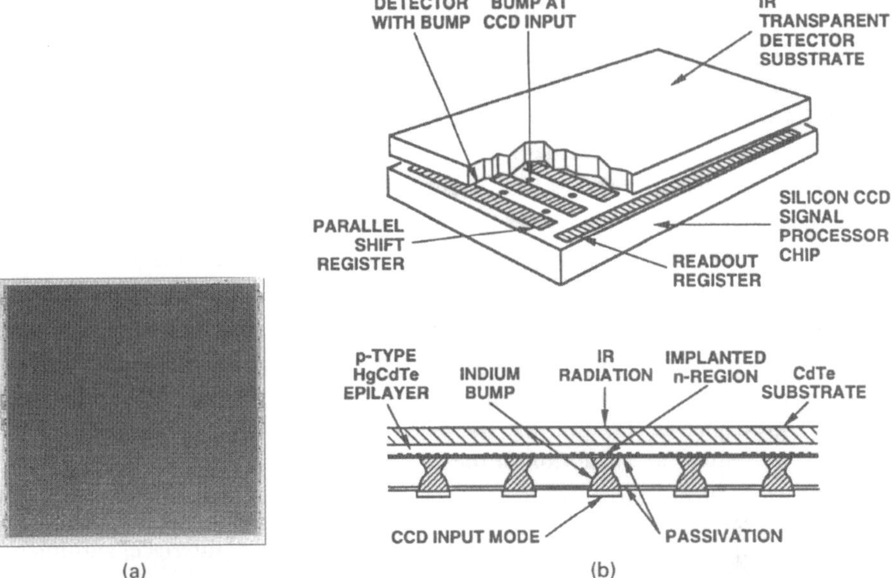

Figure 5.1 (a) A 128 × 128 pixel HgCdTe focal plane array.[8] (b) Schematic diagram of a second-generation hybrid focal plane array, consisting of a two-dimensional array of HgCdTe photodiodes connected with indium bumps to a two-dimensional signal multiplexer fabricated in silicon.[9]

Photovoltaic detectors make use of a built-in potential difference such as that produced by a $p–n$ junction. The built-in field causes photo-generated minority carriers to cross the junction, producing a current in the external circuit. Several configurations operate on the photovoltaic effect, including $p–n$ homojunctions, $p–n$ heterojunctions, and Schottky barriers. Each of these different types of devices has certain advantages for infrared detection, depending on the particular application.

Recently, more interest has been focused on $p–n$ junction photodiodes for use in combination with silicon charge coupled devices (CCD's) in hybrid focal plane arrays for direct detection in the 3–5 and 8–12 μm spectral regions. A typical hybrid design[9] is illustrated in Figure 5.1b. In this application, photodiodes are preferred over photoconductors due to their relatively high impedance, which is appropriate for coupling to the input stage of the CCD, and for their suitability for two dimensional arrays. Further, the photodiodes have a faster response and lower power dissipation than photoconductors.

Various growth techniques such as liquid phase epitaxy (LPE), metalorganic chemical vapor deposition (MOCVD), molecular beam epitaxy (MBE), and metalorganic molecular beam epitaxy[17] (MOMBE) have been researched and developed to grow a number of detector structures including PV double layer heterojunction devices.[18, 19] By far, LPE[20, 21] is the most developed and mature growth technique for focal plane array applications. However, progress on MBE[18, 22–25] and MOCVD[26] over the last four or five years has shown that vapor phase growth techniques, because of better control over layer thickness and compositional grading, will become important for the growth of advanced device structures that require several thin layers with sharp interfaces such as two-color detectors, to meet both military and commercial requirements.

Although several laboratories have achieved significant advances toward making HgCdTe detector systems practical, reliable, and manufacturable, some problems remain. Reliable procedures for growing large area substrates, substrate cleaning, epitaxial layer growth with controlled doping, surface passivation, metallization and junction formation have been developed, but problems persist in the yield, uniformity, and reproducibility of HgCdTe detector arrays. Although the work has benefited greatly from the extensive use of characterization techniques, additional progress will require an expansion of the range of techniques that are applied.

Representative Device Structure

The purpose of this chapter is to illustrate the use of surface, interface, and microanalysis techniques to solve several practical problems associated with the fabrication of HgCdTe focal plane arrays. These techniques provide information that can be applied to aid the manufacturing yield, improve device performance, and even identify the failure mechanisms. Figure 5.2 shows a schematic cross section of a HgCdTe double layer heterojunction mesa diode structure with five important interfaces encountered during the fabrication of PV detectors. In this configuration the infrared radiation enters through the substrate and buffer layer, which are transparent at the wavelengths of interest, and is absorbed in the base layer, producing minority carriers that cross the cap/base junction to produce the output signal. The cap/base structure can be either p-on-n or n-on-p. The buffer layer is often used in vapor phase growth (MBE or MOCVD), but is omitted in LPE growth. Control of these layers and the interfaces between them is crucial for proper yield and performance. In this chapter, we will follow this drawing in order to trace the fabrication process step by step. We will simultaneously show how surface and microanalysis characterization techniques can be applied for monitoring and improvement.

5.2 Substrates and the CdTe Surface (Interface 1)

Substrate Quality

CdTe is not only used extensively for infrared detector applications, but is also important in the area of optoelectronics, solar energy conversion, and gamma ray detection.[27–29] CdZnTe is now replacing CdTe as the substrate for epitaxial growth of HgCdTe, because it can be lattice matched by adjusting the ZnTe mole fraction, and because it has greater mechanical hardness and lower dislocation density. For the epitaxial growth of semiconductors the quality of the substrate and the method

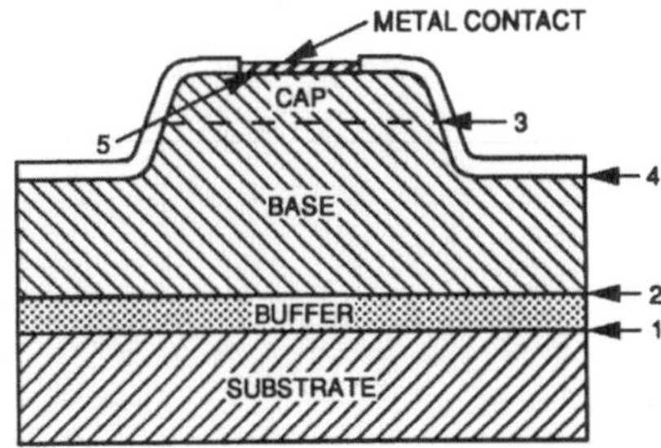

Figure 5.2   A schematic drawing of a HgCdTe double-layer heterojunction mesa diode structure with five important interfaces encountered during the fabrication of PV detectors.

of substrate preparation before growth are crucial factors determining the quality of the epitaxy. Because crystalline defects in the substrate can extend into the epilayer, it is necessary to select substrates with a minimum of dislocations and other defects such as precipitates. Impurities present in the substrates can also diffuse into the epilayers during high temperature growth, leading to unintentional doping.[30, 31]

To be acceptable for the growth of device-quality HgCdTe, the CdZnTe substrate must first be shown to have a relatively low dislocation density. For (111) oriented material this is determined by means of the Nakagawa etch,[32] which produces an etch pit wherever a dislocation intersects the surface. Good material will have an average etch pit density (EPD) of mid-10^4 to mid-10^5 cm^{-2}, except near the surface of the boule where it is higher.

X-ray topography and rocking curve techniques have been widely used to examine the crystalline quality of bulk substrates and are often used for screening substrates. A good substrate should be free of mosaic structure as seen by X-ray topography[33, 34] and would have a 10 to 15 arc-s width (full width at half maximum, FWHM) of the X-ray rocking curve. The technique of synchrotron white beam X ray topography (SWBXT) is extremely sensitive to strains and structural defects.[34] In addition, transmission electron microscopy[35, 36] (TEM), photolumines cence[37] (PL), and secondary ion mass spectrometry[38] (SIMS) are often used to determine the microstructural defects, precipitates, and impurities. For instance, PL is a very sensitive optical technique that can detect the changes associated with thermal annealing treatments such as those used to alter the stoichiometry of CdTe or CdZnTe by creation of vacancies and interstitials.[37] Impurity concentrations can be measured down to the parts per billion range in bulk specimens using glow discharge mass spectrometry[39] (GDMS) and in profiles several micrometers deep by means of resonance ion spectrometry[40, 41] (RIS).

Substrate Surface Preparation

The focus of this section is on surface preparation prior to epitaxial growth. In general, degreasing, etching and preheating procedures are required before MBE

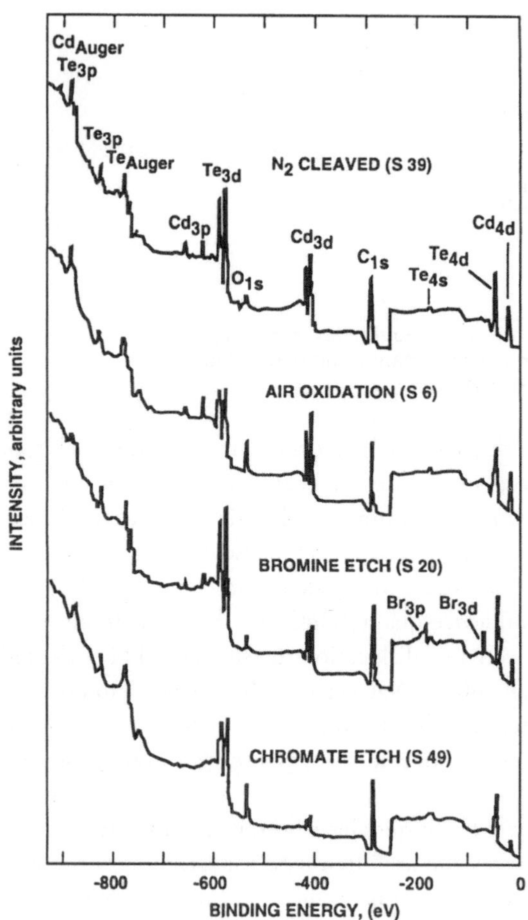

Figure 5.3   Typical XPS surveys of CdTe surfaces after the four treatments: N$_2$ cleaved (S39), air oxidation (S6), bromine etch (S20), and chromate etch (S49).[43]

and MOCVD growth, to ensure that the substrate surface is free of contaminants and oxides and is structurally ordered. Any etchant influences the chemical composition of the CdTe surface.[42] For instance, nitric acid, EDTA, and K$_2$Cr$_2$O$_7$/Ag$^+$ all leave the surface depleted of cadmium over at least some tens of angstroms and should be avoided in device fabrication. The best etchants with regard to maintaining surface stoichiometry are Br$_2$ in methanol or ethylene glycol. To prepare for MBE growth, however, it is best to follow the etches by vacuum-heating the surface, which seems to leave it stoichiometric with an extremely low sticking coefficient for oxygen. Cleaving, although impractical for device fabrication, usually leaves a stoichiometric surface, whereas air exposure leads to the formation of TeO$_2$. Typical XPS surveys of CdTe surfaces after the four treatments are shown in Figure 5.3.[43] The etchant used most widely in the community is bromine-in-methanol which

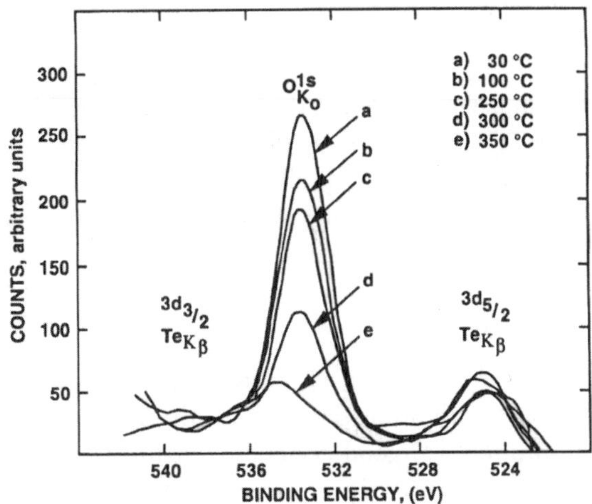

Figure 5.4   The oxygen 1s XPS spectra are shown for substrates that have been preheated at 30, 100, 250, 300, and 350 °C.[44]

leaves a very thin Te-enriched surface layer. Hydrogen heat treatment (commonly used in an MOCVD system) restores the stoichiometry on an etched CdTe surface.

The most common method for cleaning II–VI substrates prior to vapor phase growth is heat treatment under vacuum. The surface stoichiometry and oxygen contamination of the CdTe substrate surface as a function of preheating temperature in the range from 100 °C to 450 °C have been studied by Waag et al.[44] with XPS. They found that for preheating temperatures up to 200 °C the Cd/Te ratio increases steadily from about 35/65 up to nearly the ideal value of 50/50. In this temperature region, the excess Te left from the bromine-methanol etch evaporates until the surface nearly reaches the ideal stoichiometry. For preheating temperatures above 200 °C the CdTe surface becomes tellurium rich again and the higher the preheating temperature, the higher the tellurium content of the surface. For example, in Figure 5.4 the oxygen 1s XPS spectra are plotted for substrates that have been preheated at various temperatures. The oxygen signal remains nearly unchanged for temperatures below 250 °C, but it starts to decrease for preheating temperatures above 250 °C, with a steep decrease between 250 and 350 °C. For preheating temperatures higher than 350 °C, the oxygen contamination level drops below the detection limit. However, low levels of carbon and sodium can often be detected on the surface by SIMS even after the preheating treatment.

To avoid the detrimental effects of these impurities and improve the crystal surface quality a buffer layer of CdTe or CdZnTe a few micrometers thick is sometimes grown prior to the active layer growth in the MBE[23] and MOCVD techniques. This buffer layer reduces the propagation of impurities and dislocations into the active layer. For the LPE technique it is usually not necessary because the surface of the substrate melts back prior to growth.

5.3 Epitaxial HgCdTe Materials (Between Interfaces 2 and 5)

After the substrate is conditioned for epitaxial growth and the buffer layer deposition is completed, two active layers of HgCdTe are grown, as illustrated in Figure 5.2. In the p-on-n structure, the first is an n-type base layer, about 10 to 30 μm thick. The second is a p-type cap layer, which is about 1 to 5 μm thick. If the composition (x-value) differs between the two layers, their interface forms a heterojunction.

Desired Characteristics of the Active Layers

The desired properties for a base layer HgCdTe include: (1) correct HgCdTe composition, which determines the cutoff wavelength of the detector, (2) thickness greater than 10 μm for efficient photon absorption, (3) good crystalline quality and morphology, and (4) less than mid-10^{15} cm^{-3} n-type doping concentration for longer minority carrier lifetime. On the other hand, the desired properties for the HgCdTe cap layer are: (1) slightly higher Cd composition (wider bandgap) than the base layer, (2) at least 1 μm in thickness, (3) good crystalline quality and morphology, and (4) higher than 10^{17} cm^{-3} p-type carrier concentration to facilitate ohmic contacting to the external circuit. The following are the most commonly employed techniques and examples for the characterization of the HgCdTe layers.

Composition

The composition of the HgCdTe epilayer[45] can be determined by energy dispersive X-ray (EDX) analysis, optical reflectance spectroscopy, and infrared absorption and transmission technique. EDX analysis gives an average composition to 1–2 μm in depth depending on the primary beam energy (10–20 keV) used. In contrast, optical reflectance spectroscopy analyzes only the surface composition of a few hundred angstroms. Therefore, it is important to remove the surface contaminants prior to this kind of measurement to avoid surface effects. Room temperature infrared

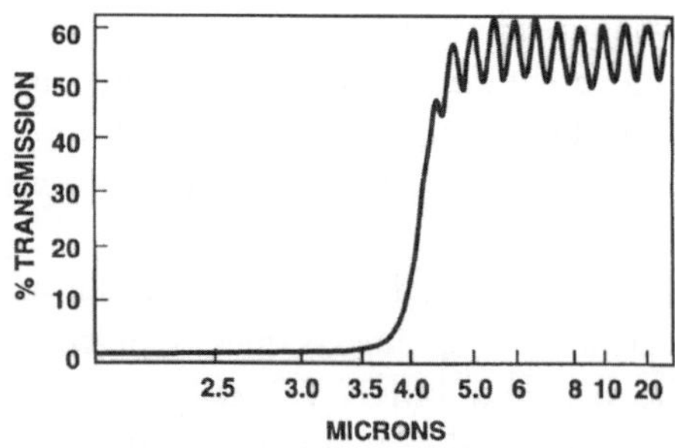

Figure 5.5 IR transmission spectrum for an MBE-grown HgCdTe layer 5 μm in thickness having a cutoff wavelength of 4 μm.

  OTHER COMPOUND SEMICONDUCTOR FILMS Chapter 5

transmission and absorption are widely used to determine the cutoff of the HgCdTe materials. For example, Figure 5.5 shows an IR transmission spectrum for an MBE-grown HgCdTe layer 5 µm in thickness having a cutoff wavelength of 4 µm. The excellent quality of the epilayer/substrate interface is demonstrated by the sharp interference fringes. The data also show that the room temperature cutoff is extremely abrupt, which is indicative of high degree in-depth compositional uniformity. Hougen has shown that the composition profile and relative photoconductive response of a detector can be predicted using room temperature transmission data alone,[46] and, of course, compositional uniformity is very crucial for the fabrication of large arrays. Usually, the compositional uniformity is assessed at multiple points on a wafer by measuring room temperature transmission and absorption, because this is nondestructive and relatively fast.

Crystalline Quality

The crystal quality of HgCdTe epilayers can be examined by the double crystal X-ray diffraction rocking curve (DCRC) technique, etch pit density, and TEM analysis. The growth of HgCdTe epilayers with extremely high structural perfection by MBE process has been demonstrated at Hughes and elsewhere.[24, 47–50] Figure 5.6a shows a DCRC measurement for a *p*-on-*n* LWIR double layer heterojunction structure. The In-doped *n*-type (about 8 µm) base layer peak has a width of 25 arc-s and is indistinguishable from the CdZnTe substrate, to which it is exactly lattice-matched. Because the As-doped *p*-type cap layer is much thinner (about 2 µm) and

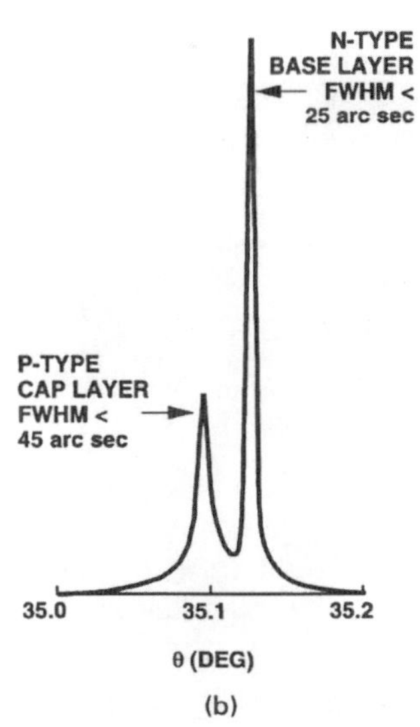

Figure 5.6   X-ray rocking curve data for a HgCdTe DLHJ structure, showing high crystallinity and uniformity of base and cap layers grown.[24]

has a different alloy composition, its peak is shifted slightly and is broader (45 arc-s), but the X-ray FWHM width is still indicative of high quality. Figure 5.6*b* shows the X-ray FWHM data map for 9 spots across the 2.5 × 2.5 cm specimen. The results indicate that the uniformity of crystallinity is excellent for focal plane applications. This represents outstanding structural perfection comparable to that obtained for high quality GaAs.

The etch pit density (EPD), to reveal the density of dislocations in (111) HgCdTe, can be determined by applying the Hähnert and Schenk etch.[51] Values of $1–5 × 10^5$ cm^{-2} are typical in good material. However, because a few micrometers of material must be sacrificed, this should be considered a destructive test for epitaxial layers, so it is normally done on a small piece cut from the wafer for this purpose. (This contrasts with EPD measurements in substrates, where the CdZnTe can simply be repolished prior to the next process step.)

Although preparation of good TEM samples of HgCdTe is technically very challenging, much useful information has been obtained about defects in the epilayer as well as imperfections at the interfaces.[35, 36] Figure 5.7 shows examples of TEM cross sections of MBE HgCdTe grown on (111) and (211) orientations.[23] The results indicate that the density of microtwins on the (111) orientation is much higher than that for the (211) orientation. This result is consistent with the DCRC measurement which indicates that the (111) orientation has an FWHM which is at least 50% larger as compared to the (211) orientation. Because of lower microtwin density and better morphology, (211) has become the preferred orientation for MBE growth.[24] For comparison, it should be noted that in LPE, where (111) is preferred, microtwins do not occur.

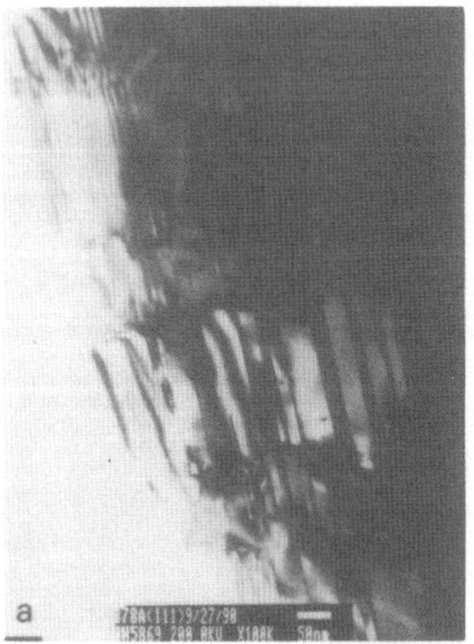
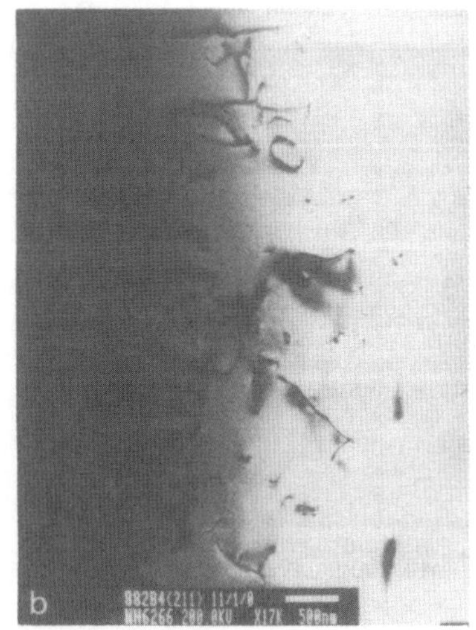

Figure 5.7   TEM cross section of MBE-grown HgCdTe in (111) and (211) orientations.[23]

Doping

The control of intentional and unintentional doping in the epilayer is another
important factor that can affect the device performance. SIMS has been used to
analyze impurities in epitaxial layers of HgCdTe because of its sensitivity to trace
impurities and because it permits profiling of intentional and unintentional doping
impurity concentrations through the thickness of the epitaxial layer. SIMS impurity
profiles are quantified using relative sensitivity factors derived from standard reference samples that have been ion implanted with known amounts of the elements
in question. SIMS has also been used to determine diffusion profiles, to determine
the change in an implant profile before and after laser annealing, and to measure
segregation coefficients for dopants in the epilayers. Several relative sensitivity factors, as well as approximate detection limits, in HgCdTe have been studied carefully by Lapides[52] and by Wilson.[38] For instance, it was found[52] that by analyzing
$TeAs^-$, a factor of 10 in sensitivity is gained over analyzing As^-; it was also found
that detection limits for In (*n*-type dopant) and As (*p*-type dopant) are 10^{13} and
10^{16} cm^{-3}, respectively. However, accurate SIMS measurement of low levels of Cu,
an impurity that kills carrier lifetime as well as acting as an acceptor, is made very
difficult because of the mass interference of Cu^{2+} and Te^{2+}, which have the same
charge-to-mass ratio. This problem is overcome by the element-selective resonance
ionization employed by RIS.[40, 41]

To illustrate the application of SIMS, an example is given on indium doping of
a multilayer structure. Memory effects for indium doping in the MBE growth of
HgCdTe were initially a major concern.[53] Work at Hughes has shown that such
effects can be minimized for a broad doping range of 10^{15} to 10^{18} cm^{-3}, as shown
in Figure 5.8.[23] In this experiment, twelve layers, each 0.3 μm thick, were grown,
with the composition being $x = 0.2$ for the first six and 0.3 for the last six, as shown
in Figure 5.8*a*. The layers were alternately undoped and In-doped, with various temperatures of the In effusion cell being 500, 600, and 700 °C. Figure 5.8*b* presents the
SIMS profile of the In concentration in this structure. It is evident that, for the same
In source temperature, the dopant concentration was the same whether x was 0.2 or
0.3. Also, there was no hint of the memory effect, because a layer can be doped at 3 ×
10^{15} cm^{-3} after a previous layer had been doped to high 10^{18} cm^{-3}. This is clear from
a comparison of the low concentration at zero depth (the last material grown) with
the doping peak at about 0.4 μm. (Variations in the other minima corresponding to
the undoped layers are due to sputter-induced roughness in the SIMS profile.)

The use of intentional dopants such as arsenic for *p*-type doping of HgCdTe is
a critical issue for successful formation of devices by MBE. Two approaches for As
doping have been investigated. The first is based upon the fact that high levels of
p-type As doping can be achieved in the binary end compound CdTe by means of
photo-assisted MBE.[47, 54] A compositionally modulated structure is grown in which
CdTe and HgCdTe layers alternate, with only the CdTe being doped. Since the CdTe
contains relatively few cation vacancies and is grown under Cd-rich conditions, the

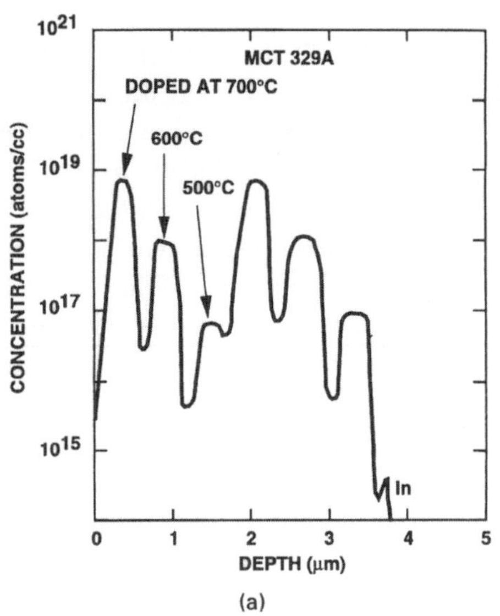

HgCdTe (x=0.3, 0.3 μm, UNDOPED)
HgCdTe (x=0.3, 0.3 μm, In=700°C)
HgCdTe (x=0.3, 0.3 μm, UNDOPED)
HgCdTe (x=0.3, 0.3 μm, In=600°C)
HgCdTe (x=0.3, 0.3 μm, UNDOPED)
HgCdTe (x=0.3, 0.3 μm, In=500°C)
HgCdTe (x=0.2, 0.3 μm, UNDOPED)
HgCdTe (x=0.2, 0.3 μm, In=700°C)
HgCdTe (x=0.2, 0.3 μm, UNDOPED)
HgCdTe (x=0.2, 0.3 μm, In=600°C)
HgCdTe (x=0.2, 0.3 μm, UNDOPED)
HgCdTe (x=0.2, 0.3 μm, In=500°C)
CdZnTe SUBSTRATE

(b)

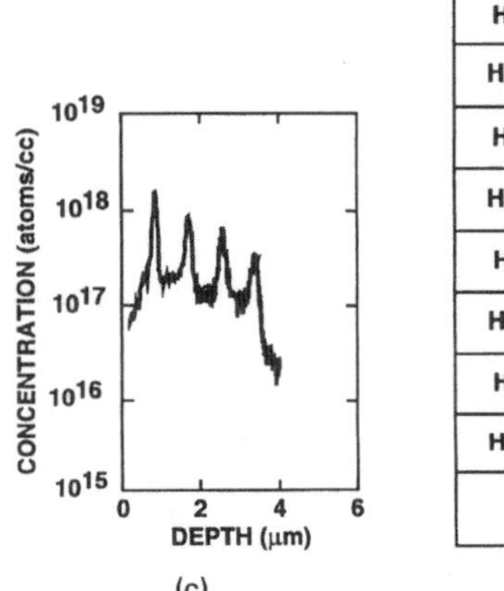

HgCdTe (x=0.35, 0.45 μm, UNDOPED)
HgCdTe (x=0.35, 0.45 μm, As=280°C)
HgCdTe (x=0.35, 0.45 μm, UNDOPED)
HgCdTe (x=0.35, 0.45 μm, As=260°C)
HgCdTe (x=0.35, 0.45 μm, UNDOPED)
HgCdTe (x=0.35, 0.45 μm, As=240°C)
HgCdTe (x=0.35, 0.45 μm, UNDOPED)
HgCdTe (x=0.35, 0.45 μm, As=220°C)
HgCdTe (x=0.35, 0.45 μm, UNDOPED)
CdZnTe SUBSTRATE

(d)

Figure 5.8 (*a* and *b*) Schematic diagram of an MBE-grown In-doped multilayer HgCdTe structure and its In SIMS profile.[23] (*c* and *d*) Schematic diagram of an MBE-grown As-doped multilayer HgCdTe structure and its As SIMS profile.[23] (*e*) SIMS profiles of As and In concentrations for an *n*-on-*p* heterojunction.[52]

As is properly incorporated on the anion sites in proportion to the As flux. The structure is subsequently annealed to interdiffuse the layers, forming homogeneous *p*-type HgCdTe with efficient activation of the As. The disadvantage of this approach is that it sacrifices control of the desired junctions and interfaces, such as *p*–*n* junctions, because they also interdiffuse. The second approach[22–24] is to grow

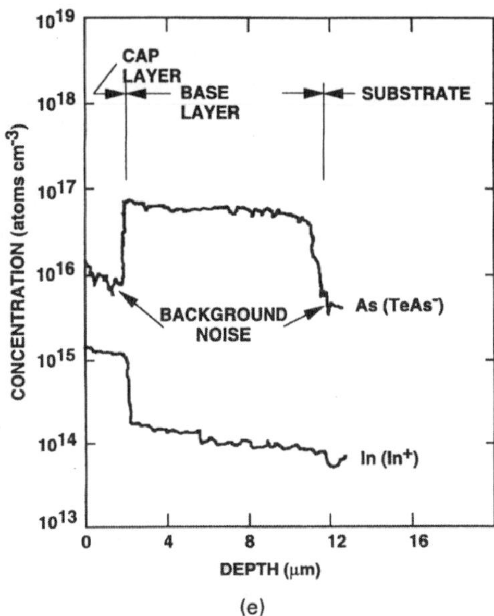

Figure 5.8 *continued*

the desired composition of HgCdTe directly, using cadmium arsenide compound and atomic tellurium from a cracker cell (instead of the conventional Te_2 molecule) to enhance the sticking coefficient of Hg and hence to minimize the Hg vacancies. Consequently, the As is directed to the group VI sublattice (anion sites) rather than the group II sublattice, producing efficient doping. Figure 5.8*c* shows a schematic diagram of such an As-doped multilayer HgCdTe structure. In this experiment, the composition was grown uniformly at $x = 0.35$, while the As effusion cell temperature was ramped up (220 °C, 240 °C, 260 °C, 280 °C) and the As effusion cell was alternately shuttered off and on. Figure 5.8*d* shows the SIMS profile of the resultant As concentration, demonstrating that As incorporation can be achieved in the range of 10^{16} to 10^{18} cm^{-3} with the help of the cracker cell on the Te source.

For the Hg-melt LPE growth method,[20, 52] Figure 5.8*e* shows profiles of As and In concentrations for an *n*-on-*p* heterojunction. Although the cap layer was grown at 370 °C, there was no significant interdiffusion of dopants from either the cap or the base layer.

Electrical properties of HgCdTe layers are characterized by Hall effect and C–V measurements. Hall effect measurements on *n*-type and highly doped *p*-type samples are relatively straightforward. However, the characterization of low doped *p*-type samples is complicated by mixed conduction effects because of the much greater mobility of electrons than holes in HgCdTe. Lou and Frye[55] have developed a computer model to extract some fundamental properties such as surface charge density, bulk carrier concentration, surface and bulk mobility, and dopant concentration in double-layer heterojunctions from the experimental data otherwise classified as "anomalous" and not previously analyzable. They have identified this kind of anomalous Hall data with three configurations, namely: n skin on p layer, n layer on p layer, and n^+ skin on n layer. The applicability of their model to LPE HgCdTe has been substantially demonstrated. Figure 5.9a shows a variable temperature Hall effect measurement on an x = 0.21 layer grown by MOCVD which had a chemically determined As level of 1×10^{16} cm^{-3} (Reference 56). Theoretical fitting has been performed using the model developed by Capper et al.[57] The best fit was obtained assuming N_a = 9 $\times 10^{15}$ cm^{-3} with a donor compensation level, N_d, of 3×10^{15} cm^{-3}. The ionization energy of the As dopant was found to be 7 meV, which is consistent with the shallow level expected from an acceptor impurity in this type of composition and carrier concentration.[23] Figure 5.9b shows the mobility and Hall coefficient data for a 3.5 μm thick MOCVD-grown HgCdTe layer with a composition of x = 0.3, grown on a CdTe substrate.[58] The layer exhibits n-type behavior with a carrier concentration at a low temperature of 2×10^{15} cm^{-3} and an electron mobility of 40 000 cm^2/V·s at 40 K. The n-type behavior of this layer is attributable to the low growth temperature and impurities in the newly developed source material. However, higher mobility values of n-type In-doped HgCdTe (greater than 1×10^5 cm^2/V·s) have been reported recently by MBE.[24]

Minority Carrier Lifetime

For both the photoconductor and photodiode, the detector's responsivity and noise are related to the minority carrier lifetime and surface recombination velocity of the HgCdTe.[59, 60] The minority carrier lifetime is a measure of the bulk recombination rate of excess minority carriers and is sensitive to the HgCdTe material properties such as composition, carrier concentration and material defects. The bulk recombination lifetime is determined by Auger, radiative and Shockley–Read–Hall recombination processes. The Shockley–Read–Hall recombination is associated with defect states within the bandgap. However, the Auger and radiative recombination rates are determined by the electronics band structure of the semiconductor.

In general, lifetime is either measured directly or extracted from the measured value of another semiconductor parameter, such as carrier diffusion length. Lifetime measurements are based on monitoring the time evolution of an excess carrier density produced by carrier injection methods, such as photon, electron beams, X rays or electronic fields. Photoconductivity (PC) decay measurement is the most used technique to determine the minority carrier lifetime in HgCdTe materials. Typically, the conductivity decay is fitted to an exponential function and the lifetime is defined as the time required for the conductivity to decay to $1/e$ of its maximum value. Recently, Wijewarnasuriya et al.[60] have studied the recombination mechanism of In-doped MBE grown HgCdTe epilayers. Measured minority carrier lifetime can be explained by an Auger limited band-to band recombination process in this material. They have observed lifetimes of the order of 500–800 ns

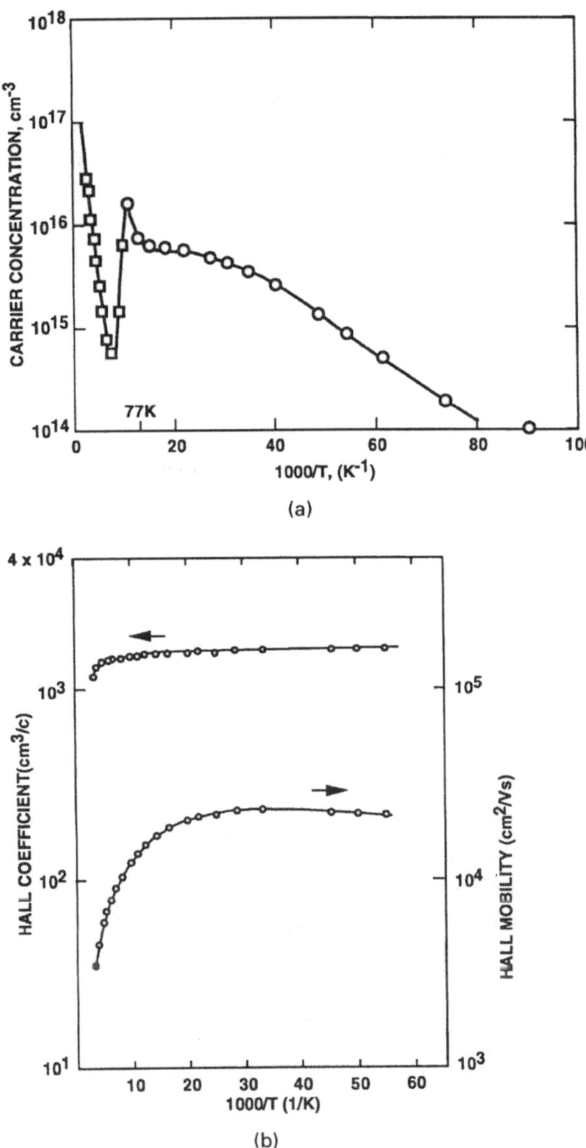

Figure 5.9   (a) A variable-temperature Hall effect measurement on an $x = 0.21$ layer grown by MOCVD which had a chemically determined As level of 1×10^{16} cm^{-3}. (From Reference 56.) (b) The mobility and Hall coefficient data for a 3.5 µm thick MOCVD-grown undoped HgCdTe layer with a composition of $x = 0.3$ grown on a CdTe substrate.[58]

at 80 K. The carrier concentration and mobility at 80 K in these layers were typically 2×10^{15} cm^{-3} and about 1×10^5 cm^2/V·s, respectively for HgCdTe with $x = 0.22$ to 0.23.

5.4 Heterojunction Interfaces (Interface 3)

Advantages of the Heterojunction

The heterojunction interface is the most crucial interface for a double layer p–n heterojunction PV detector. For example, a properly constructed heterojunction detector offers improved reverse breakdown characteristics at very low temperatures and long detection wavelengths due to reduced tunneling current. At higher operating temperatures, diode leakage due to diffusion from the wide bandgap region is eliminated.

Desired Characteristics

A potential issue is the formation of a barrier to the transport of photogenerated minority carriers across the junction. To avoid heterojunction barriers requires careful control of parameters such as the doping, the composition difference, and most importantly the compositional grading across the p–n heterojunction. Bratt[62] showed that the relative position and width of the graded bandgap region (metallurgical junction) within the depletion region of the p–n junction was the dominant factor in the formation of barriers. His general prescription for the avoidance of barriers is that: (1) the width of the graded heterojunction (metallurgical junction) should be greater than about 0.2 µm to avoid discontinuities of the type predicted by the Anderson model and (2) either the graded bandgap region should be entirely contained within the depletion region or the p–n junction should be located on the narrow-gap side of the heterojunction. In fact, placement of the graded gap region entirely within the depletion region of the p–n junction is preferable to pushing the electrical junction into the narrow gap region. This is because the beneficial effects of the wider energy gap in reducing generation-recombination and tunneling leakage currents are lost if the p–n junction occurs in the low bandgap region. That is, if the depletion layer is near the center or toward the wide

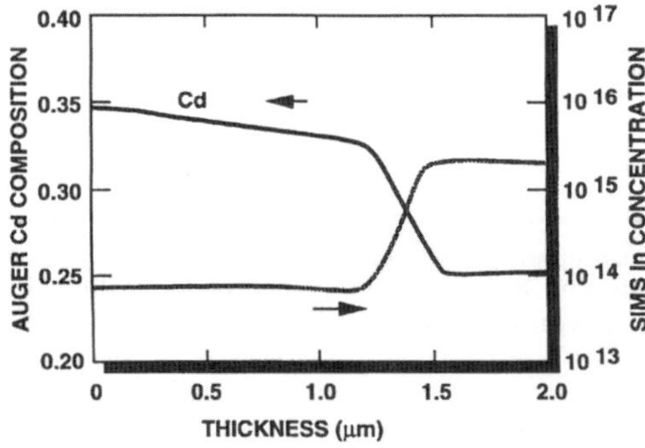

Figure 5.10   Auger and SIMS depth profiles of a *p*-on-*n* DLHJ structure showing the metallurgical and electrical junction location at the graded region.

bandgap side of the transition region, then a potential barrier is formed which impedes the flow of electrons across the heterojunction. This results in photodiodes with a high dynamic resistance but poor minority carrier collection efficiency. When the depletion layer is located near the narrow bandgap side of the hetero-junction, however, the potential barrier goes away and normal photodiode characteristics are observed.

Characterizations

The graded gap region (metallurgical junction) can be characterized by the Auger electron spectroscopy (AES) depth profiling of Cd as shown in Figures 5.10 and Figure 5.11. One curve in Figure 5.10 shows the Auger depth profile of Cd for an MBE-grown double-layer *p*-on-*n* heterojunction. The graded region is estimated to be about 0.3 µm wide. To complement the Auger analysis, the location of the *p*–*n* electrical junction can be determined by SIMS. The other curve in Figure 5.10 shows the In doping profile of the same structure obtained by SIMS, using argon as the primary ion. The result indicates that the base layer has an In concentration of 3×10^{15} cm^{-3} and the *p*–*n* junction is located about at the center of the graded region, which is consistent with the MBE growth conditions. In contrast, Figure 5.11 shows the AES and SIMS profiles obtained from a poorly performing diode, having a graded region less than 0.1 µm and the *p*–*n* junction located on the low bandgap side. These profiles are consistent with the MBE growth parameters, but the graded region is too abrupt for optimal performance. Also, device diagnostics in general have been helped by a novel use of SIMS, in which the *x*-value can be measured by recording the Te signal obtained with a Cs$^+$ beam, because it has been found to be proportional to the CdTe mole fraction of the HgCdTe.[63]

Electron beam induced current (EBIC) imaging is a useful scanning electron microscope based technique for analysis of semiconductors, which has been extensively applied to silicon *p*–*n* junctions, Schottky barriers, and MIS capacitors for

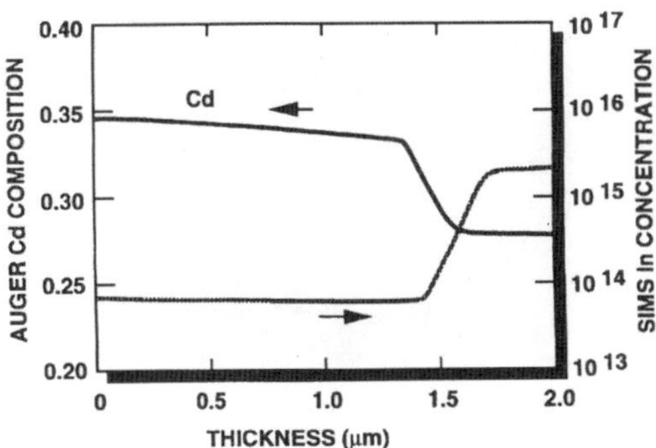

Figure 5.11 Auger and SIMS depth profiles of a *p*-on-*n* DLHJ structure showing that the electrical junction location is near the base layer region.

the purpose of investigating materials or processing induced defects, compositional inhomogeneities, and measuring minority-carrier diffusion lengths. Application of EBIC to HgCdTe has been less extensive, but it can provide useful information about interfaces and defects.[64] Light beam induced current (LBIC) has also proven to be a very useful diagnostic.[65] TEM analysis is another powerful technique used to examine the junction interface characteristics. Figure 5.12 shows a TEM cross-section of a poorly performing *p*-on-*n* diode grown by MBE. The TEM analysis identified Te-enriched precipitates at the metallurgical junction which are most likely the cause of the poor performance. The precipitates can be eliminated with the adjustment of Hg and Te beam fluxes.

5.5 HgCdTe Surface Preparation (Interfaces 4 and 5)

Importance of the Chemically Etched Surface

A clean-up of the HgCdTe surface, by means of chemical etching, is performed at several points in every practical process for device fabrication. Typically, cleaning immediately precedes metallization for contact purposes, the growth of a native oxide or native sulfide for passivation, the deposition of a dielectric layer as a passivation or antireflection coating, and sometimes, the growth of another epitaxial layer. Cleaning is also performed prior to material characterizations, such as Hall effect

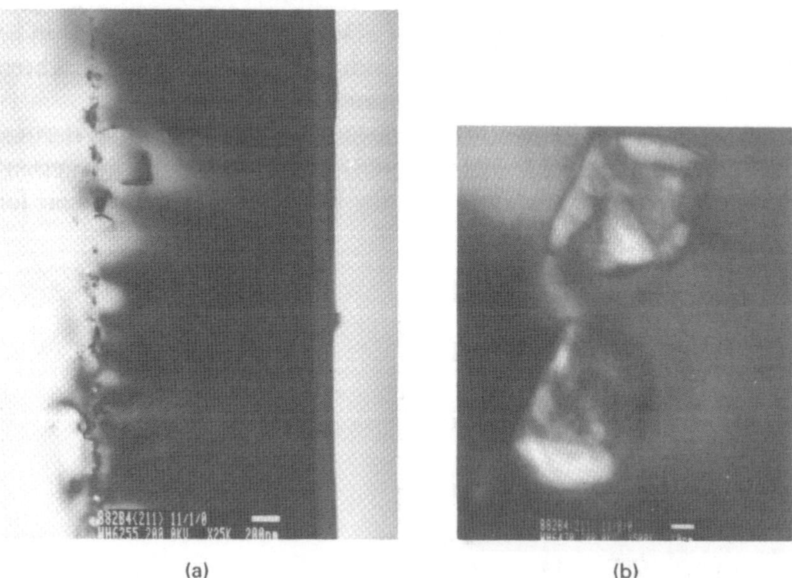

(a)   (b)

Figure 5.12　(*a*) Low-magnification TEM image of MBE-grown HgCdTe *p*-on-*n* heterostructure showing Te precipitates at the junction. (*b*) High-magnification TEM of Te precipitates. Typical size is about 500 Å.

and optical reflectance, the results of which are sensitive to the condition of the surface.

The goal of the clean-up treatment is to establish a new and reproducible surface, so that a process or characterization will not be affected by prior exposure to ambient conditions or other process substances, such as photoresist. Throughout the industry, the most common etchant is a solution of Br_2 in a solvent such as methanol or ethylene glycol.

A "clean" chemically etched surface, however, can actually exhibit carbon compounds, air oxidation, nonstoichiometry, lattice disorder, and physical roughness, as one can demonstrate with a variety of analytical techniques. In any attempt to develop or monitor a device fabrication process, the challenge is to understand the nature of this surface and to achieve a degree of control over its characteristics. Below we give examples of analyses applied to this purpose.

Monitoring of the Surface Cleanliness by Ellipsometry

Ellipsometry is a sensitive, rapid, and nondestructive means of characterizing surfaces and films. The two principal applications of ellipsometry to HgCdTe[66] are the characterization of (1) films grown or deposited with thicknesses in the range of 100 to more than 5000 Å, and (2) surfaces having process residues or deliberate films thinner than about 100 Å. The former will not be discussed here because the approach is the same as for other semiconductors. Instead we will discuss the aspects of the latter that are peculiar to HgCdTe.

The interpretation of ellipsometric data from very thin (≤ 100 Å), and often unknown, films on HgCdTe is best done in a semiquantitative manner, by graphically comparing each reading with theoretical plots of known films. The procedure is illustrated in Figure 5.13. We assume standard conditions of a 6328 Å wavelength (He-Ne laser) and a 70° angle of incidence, measured from the normal to the surface. Each ellipsometric reading is expressed in terms of the two parameters ψ and Δ, which are confined to the range $0° \leq \psi \leq 90°$ and $0° \leq \Delta \leq 360°$. Figure 5.13 covers only the small region of the ψ–Δ plane surrounding the ellipsometry reading that corresponds to clean, smooth $Hg_{0.7}Cd_{0.3}Te$, as indicated. Emanating from this point are curves of the theoretical readings for films on the surface of $Hg_{0.7}Cd_{0.3}Te$, based upon the refractive index n, extinction coefficient k, and thickness d of the film material. Native oxides are represented by the two curves of $n = 2.0$ and 2.5, with $k = 0$. In general all films having $k = 0$, including typical organic residues, will cause Δ to fall below 148° by approximately 1 degree per 3.5 angstroms of thickness. Microroughness can be defined as the presence of numerous topographical features that are small compared to the wavelength of the probing light. Microroughness also produces a drop in Δ but cannot easily be quantified. The curve for films of amorphous Te is shown in the figure. Also, curves for amorphous Te mixed with 30% TeO_2 and amorphous Te mixed with 30% void spaces are shown. Thus Te and its typical mixtures can produce a rise in Δ, and a

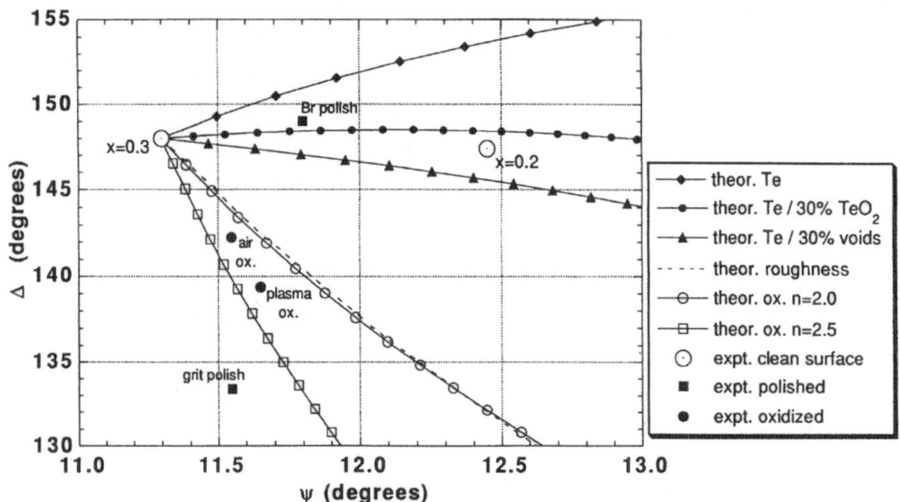

Figure 5.13 **Theoretical and experimental ellipsometry readings for very thin films on HgCdTe. Curves represent increasing film thicknesses, with symbols located at intervals of 5 Å. All features apply to Hg$_{1-x}$Cd$_x$Te with x = 0.3, except for the single point indicating the clean surface for x = 0.2. The extreme sensitivity of ellipsometry is evident.**

substantial shift of ψ to the right. Although only one curve for each kind of mixture is shown, theoretically, as the Te fraction in these mixtures decreases, their curves would swing downward. Finally some typical experimental measurements are presented for comparison.

The unambiguous analysis of ellipsometric readings in the region covered by Figure 5.13 requires a knowledge of the sample history, aided by other surface analytical techniques such as XPS. Nevertheless, one can easily monitor a routine surface cleaning process by using ellipsometry alone to determine whether a surface has been properly cleaned. To do this, one sets an acceptance window in the ψ–Δ plane, such as $11.1° \leq \psi \leq 11.5°$ with $147° \leq \Delta \leq 149°$. Similarly, one could set another acceptance window to monitor a process that grows or deposits a thin native oxide or other dielectric layer.

For metal films, unfortunately, ellipsometry is relatively insensitive when the thickness exceeds 40 or 50 Å, because metals typically have a large extinction coefficient k, which limits the penetration of the light.

The ideal reading for clean, smooth Hg$_{0.8}$Cd$_{0.2}$Te is also shown in Figure 5.13. In this region of x, the ψ of the substrate increases by 1.1° when x drops by 0.1.

Characterization of Thin Native Oxides on HgCdTe by XPS

XPS has proven to be very effective in characterizing the HgCdTe surface, in part because of good quantification, but especially because of its sensitivity to oxidation states. Figure 5.14 shows the results of XPS sputter profiling of LPE-grown

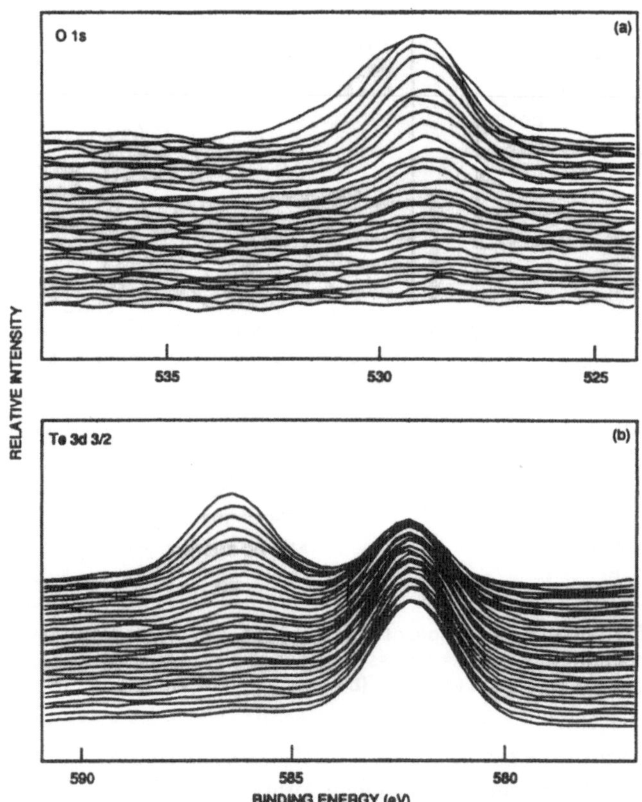

Figure 5.14 High resolution XPS spectra of a $Hg_{0.8}Cd_{0.2}Te$ surface. Curves in each panel are separated by the sputter removal of about 3 Å of material, with the outermost curve being at the top. (*a*) The O 1*s* peak, which diminishes with depth. (*b*) The Te $3d_{3/2}$ peak, showing a chemical splitting.

$Hg_{0.8}Cd_{0.2}Te$ after spray etching with a solution of Br_2 in ethylene glycol, followed by 48 h of air oxidation. In thirty steps, spectra were obtained alternating with sputtering by 4 keV Ar^+ ions, reaching a total depth of 90 Å. The elements monitored at high resolution were Hg, Cd, Te, O, and C. The resulting sequences of spectra for oxygen and Te are displayed in Figure 5.14. The Te $3d_{3/2}$ transition was chosen rather than the more intense Te $3d_{5/2}$, in order to avoid a possible interference with the Hg $4p_{3/2}$ line. The uppermost trace in each case represents the initial surface.

The O 1*s* peak in Figure 5.14*a* exhibits a major component at 529.0 eV binding energy, which is typical of metal oxides, and a minor component at 530.7 eV. The latter is gone after about 10 Å, whereas the former persists to more than 35 Å. The Te in Figure 5.14*b* exhibits a peak at 586.5 eV, representing Te^{4+} as in TeO_2, which, as with the major oxygen peak, persists to more than 35 Å. As this dies out, another Te peak at 582.3 eV takes its place, representing the telluride phase.

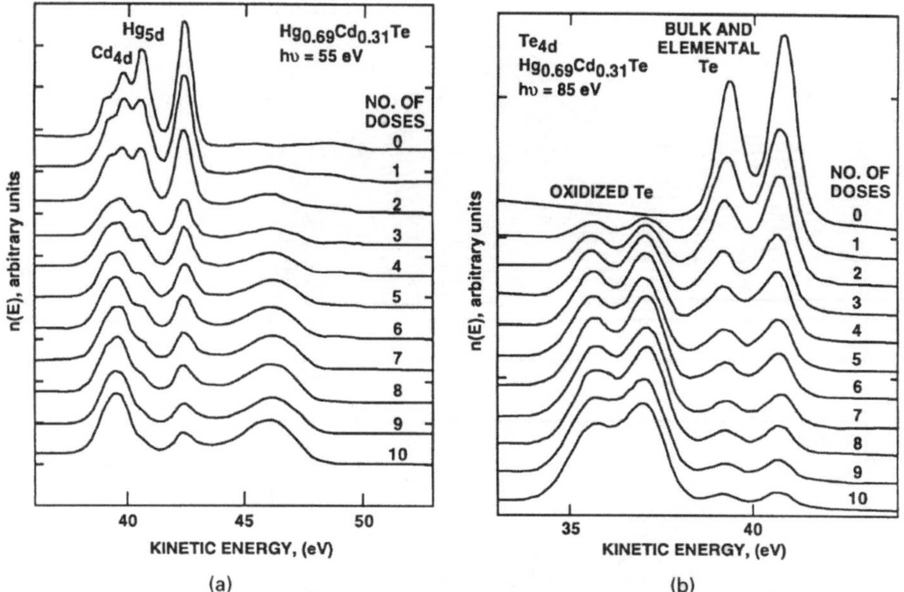

Figure 5.15 UPS spectra of the surface of a sample of $Hg_{0.69}Cd_{0.31}Te$. The sample was cleaved in vacuum and exposed to repeated doses of activated oxygen. (*a*) Peaks due to Cd 4*d*, Hg 5*d*, and valence band emission. (*b*) Te 4*d* peaks.

In summary, the chemical shifts for Te are very clearly resolved, and the "oxide" state of Te correlates very well with the oxygen signal. Although the depth information is convoluted with an electron escape depth of about 20 Å, the native oxide thickness may be estimated from the half-height point versus depth to be about 30 Å. The minor oxygen peak may represent another compound in the native oxide, or an organic contaminant.

Surface Analysis by UPS

For fundamental investigations of reactions involving oxygen or metals on the HgCdTe surface, ultraviolet (also called soft X-ray) photoelectron spectroscopy (UPS) has proven to be invaluable. Using a lower photon energy than XPS, the UPS technique offers greater surface selectivity due to a shorter electron escape depth, and permits a higher resolution examination of electron states near the Fermi level due to a smaller kinetic energy of the emitted electrons. The most commonly used photon energies are 21.1 and 40.8 eV, conveniently obtained from a He discharge lamp. Alternatively, one may use synchrotron radiation.

Figure 5.15 presents the results of a study[67] of the uptake of oxygen on the (110) cleaved face of $Hg_{0.69}Cd_{0.31}Te$. UPS spectra were obtained with synchrotron-generated photons of 55 and 85 eV after successive oxidizing doses produced by exposing the surface to 5×10^{-5} torr of oxygen excited by a hot filament. (No sputtering is involved.) As the oxidation proceeds, Figure 5.15*a* shows a peak due to O 2*p* growing at 46 eV, while the Hg $5d_{5/2}$ and $5d_{3/2}$ lines are attenuated and the Cd $4d$ structure is essentially unchanged. Figure 5.15b displays the attenuation of the Te $4d$ doublet representing bulk and elemental Te, while another Te $4d$ doublet shifted by 3.6 eV, representing oxidized Te, is growing, Careful analysis of the Te trace after one oxygen dose reveals that, in addition to the telluride and oxide states, a small amount of the Te in the elemental state is present.

Conclusions from this study are that both Te and Cd participate in the oxide layer, but Hg does not. It presumably goes to the vapor phase as a result of the breaking of the Hg-Te bond due to the energy carried by the excited oxygen. This explanation is consistent with the presence of elemental Te, and is supported by a similar experiment on the binary compound HgTe. Furthermore, when a cleaved HgCdTe surface is exposed to oxygen with no excitation, it does not change.[68] Thus, the ejection of Hg is a result of the energy carried by the excited oxygen and is a necessary precursor to the growth of the oxide layer.

5.6 Summary

In this chapter, we have chosen the HgCdTe p-on-n double layer heterojunction structure shown in Figure 5.2 to illustrate the five important interfaces encountered during the fabrication of PV detectors. We also simultaneously show how surface and microanalysis characterization techniques can be applied for monitoring and improvement of the processes. We present a multi-analysis approach which includes structural (X-ray, TEM, and EPD), chemical (SIMS, GDMS, RIS, EDX, Auger, XPS, and UPS), electrical (Hall, C–V, EBIC, LBIC, and PC decay), and optical (PL, FTIR, and ellipsometry) characterizations. The techniques complement one another, providing more insight than any isolated analysis. Each specific technique is chosen to provide information required for better understanding of the surface, interface, and bulk material properties as they relate to the device performance. This multi-analysis approach is often effective to enhance the communication between the material growth scientists and device engineers.

References

1   *Mercury Cadmium Telluride.* Semiconductors and Semimetals, Vol. 18. (R. Willardson and A. Beer, Eds.) Academic Press, New York, 1981.

2   D. A. Scribner, M. R. Kruer, and J. M. Killiany. *Proc. IEEE.* **79**, 66, 1991.

3   R. Balcerak, J. Gibson, W. Gutierrez, and J. Pollard. *Optical Engr.* 26, 191, 1987.

4   P. R. Norton. *Optical Engr.* **30**, 1649, 1991.

5   A. Rogalski and J. Piotrowski. *Progr. Quantum Electron.* **12**, 87, 1988.

6   R. Dornhaus and G. Nimtz. In *Narrow Gap Semiconductors.* Springer Tracts in Modern Physics, Vol. 98. Springer, Berlin, 1983, p. 119.

7   R. A. Reynolds. *J. Vac. Sci Technol. A.* 7, 269, 1989.

8   G. A. Walter and E. L. Dereniak. *Laser Focus/Electro-Optics.* **22**, 86, 1986.

9   M. Kruer, D. Scribner, and J. Killiany. *Optical Engr.* **26**, 182, 1987.

10 R. Balcerak and L. Brown. *J. Vac. Sci. Technol. B.* **10**, 1353, 1992.

11 M. A. Herman and M. Pessa. *J. Appl. Phys.* **57**, 2671, 1985.

12 R. Triboulet, M. Bourdillot, A. Durand, and T. Nguyen Duy. *Proc. SPIE.* **1106**, 40, 1989.

13 E. A. Patten, M. H. Kalisher, G. R. Chapman, J. M. Fulton, C. Y. Huang, P. R. Norton, M. Ray, and S. Sen. *J. Vac. Sci. Technol. B.* **9**, 1746, 1991.

14 S. Sen and J. E. Stannard. *Mat. Res. Soc. Symp. Proc.* **302**, 391, 1993.

15 P.-K. Liao, J. H. Hurst, G. Westphal, J. T. Tregilgas, D. F. Weirauch, H. F. Schaake, C. K. Ard, B. E. Dean, G. T. Neugebauer, A. B. Bollong, A. Socha, R. Balasubramanian, P. W. Norton, J. P. Tower, S. Tobin, and M. Kestigian. Presented at 1993 U.S. Workshop on Mercury Cadmium Telluride, Seattle. To be published in *J. Electron. Mater.*

16 M. A. Kinch. In *Mercury Cadmium Telluride.* Semiconductors and Semimetals, Vol. 18. (R. Willardson and A. Beer, Eds.) Academic Press, New York, 1981, p. 313.

17 C. J. Summer, B. K. Wagner, and R. G. Benz II. *Proc. SPIE.* **1735**, 56, 1992.

18 J. M. Arias, J. G. Pasko, M. Zandian, S. H. Shin, G. M. Williams, L. O. Bubulac, R. E. DeWames, and W. E. Tennant. *J. Electron. Mater.* **22**, 1049, 1993.

19 W. E. Tennant, C. A. Cockrum, J. B. Gilpin, M. A. Kinch, M. B. Reine, and R. P. Ruth. *J. Vac. Sci. Technol. B.* **10**, 1359, 1992.

20 T. Tung, L. V. DeArmond, R. F. Herald, P. E. Herning, M. H. Kalisher, D. A. Olson, R. F. Risser, A. P. Stevens, and S.J. Tighe. *Proc. SPIE.* **1735**, 109, 1992.

21 C. A. Castro. *Proc. SPIE,* **2021**, 2, 1993.

22 O. K. Wu. *Mat. Res. Soc. Symp. Proc.* **302**, 423, 1993.

23 O. K. Wu, D. N. Jamba, and G. S. Kamath. *J. Crystal Growth.* **127**, 365, 1993.

24 O. K. Wu. *Proc. SPIE.* **2021**, 79, 1993.

25 T. Kanno, M. Saga, A. Kawahara, R. Oilkawa, A. Ajisawa, Y. Tomioka, S. Murashima, T. Shima, and N. Yasuda. *Proc. SPIE.* **2020**, 41, 1993.

26 S. J. C. Irvine. *Proc. SPIE.* **1735**, 92, 1992.

27 H. B. Barber, H. H. Barrett, T. S. Hickernell, D. P. Kwo, J. M. Woolfenden, G. Entine, and C. O. Baccash. *Med. Phys.* **18**, 373, 1991.

28   J. F. Butler, F. P. Doty, B. Apotovsky, J. Lajzerowicz, and L. Verger. *Materials Sci. Engr.* **B16**, 291, 1993.

29   M. Cuzin, F. Glasser, J. Lajzerowicz, F. Mathy, and L. Verger. *Proc. SPIE.* **2009**, 192, 1993.

30   D. Cooper, J. Bajaj, and R. Bowman. *Proc. SPIE.* **1106**, 88, 1989.

31   R. Korenstein, R. Olson, D. Lee, P.-K. Liao, and C. A. Castro. Presented at 1993 U.S. Workshop on Mercury Cadmium Telluride, Seattle. To be published in *J. Electron. Mater.*

32   K. Nakagawa, K. Maeda, and S. Takeuchi. *Appl. Phys. Lett.* **34**, 574, 1979.

33   T. Fanning, M. B. Lee, L. G. Casagrande, D. Dimarzio, and M. Dudley. *J. Electron. Mater.* **22**, 943, 1993.

34   M. Dudley. *Mat. Res. Soc. Symp. Proc.* **307**, 213, 1993.

35   H.-J. Kleebe, W. J. Hamilton, W. L. Ahlgren, S. M. Johnson, and M. Rühle. *Mat. Res. Soc. Symp. Proc.* **161**, 63, 1990.

36   M. Krishnamurthy, P. M. Petroff, and J. M. Arias. *J. Appl. Phys.* **73**, 7952, 1993.

37   S. Seto, A. Tanaka, Y. Masa, S. Dairaku, and M. Kawashima. *Appl. Phys. Lett.* **53**, 1524, 1988.

38   R. G. Wilson. *J. Crystal Growth.* **86**, 735, 1988.

39   A. P. Mykytiuk, P. Semeniuk, and S. Berman. *Spectrochimica Acta Rev.* **13**, 1, 1990.

40   H. F. Arlinghaus, M. T. Spaar, N. Thonnard, A. W. McMahon, T. Tanigaki, H. Shichi, and P. H. Holloway. *J. Vac. Sci. Technol. A.* **11**, 2317, 1993.

41   S. Sen, J. E. Stannard, S. M. Johnson, H. F. Arlinghaus, and G. I. Bekov. Presented at 1993 U.S. Workshop on Mercury Cadmium Telluride, Seattle. To be published in *J. Electron. Mater.*

42   M. H. Patterson and R. H. Williams. *J. Phys. D: Appl. Phys.* **11**, L83, 1978.

43   J. P. Häring, J. G. Werthen, R. H. Bube, L. Gulbrandsen, W. Jansen, and P. Luscher. *J. Vac. Sci. Technol. A.* **1**, 1469, 1983.

44   A. Waag, Y. S. Wu, R. N. Bicknell-Tassius, and G. Landwehr. *Appl. Phys. Lett.* **54**, 2662, 1989.

45   S. L. Price and P. R. Boyd. *Semicond. Sci. & Technol.* **8**, 842, 1993.

46   C.A. Hougen. *J. Appl. Phys.* **66**, 3763, 1989.

47   J. Schetzina, J. Han, Y. Lansari, N. Giles, Z. Yang, S. Hwang, and J. Cook, Jr. *J. Crystal Growth.* **101**, 23, 1990.

48   R. Yanka, K. Harris, L. Mohnkern, and T. Myers. *J. Crystal Growth.* **111**, 715, 1991.

49 K. Harris, T. Myers, R. Yanka, L. Mohnkern, R. Green, and N. Otsuka. *J. Vac. Sci. Technol. A.* **8**, 1013, 1990.

50 T. H. Myers, R. W. Yanka, K. A. Harris, A. R. Reisinger, J. Han, S. Hwang, Z. Yang, N. C. Giles, J. W. Cook, Jr., J. F. Schetzina, R. W. Green, and S. McDevitt. *J. Vac. Sci. Technol. A.* 7, 300, 1989.

51 I. Hähnert and M. Schenk. *J. Crystal Growth.* **101**, 251, 1990.

52 L. Lapides. *Surface & Interface Analys.* 7, 211, 1985.

53 M. Boukerche, P. Wijewarnasuriya, S. Sivananthan, I. Sou, Y. Kim, K. Mahavadi, and J. Faurie. *J. Vac. Sci. Technol. A.* **6**, 2830, 1988.

54 J. Arias, S. Shin, D. Cooper, M. Zandian, J. Pasko, E. Gertner, R. DeWames, and J. Singh. *J. Vac. Sci. Technol. A.* **8**, 1025, 1990.

55 L. F. Lou and W. H. Frye. *J. Appl. Phys.* **56**, 2253, 1984.

56 C. D. Maxey, P. Capper, P. A. C. Whiffin, B. C. Easton, I. Gale, and J. B. Clegg. *J. Crystal Growth.* **101**, 300, 1990.

57 P. Capper, J. J. G. Gosney, C. L. Jones, I. Kenworthy, and J. A. Roberts. *J. Crystal Growth.* **71**, 57, 1985; C. D. Maxey, P. Capper, P. A. C. Whiffin, B. C. Easton, I. Gale, and J. B. Clegg. *Materials Lett.* **8**, 385, 1989.

58 S. Ghandi, N. Taskar, K. Parat, D. Terry, and I. Bhat. *Appl. Phys. Lett.* **53**, 1641, 1988.

59 V. C. Lopes, A. J. Syllaios, and M. C. Chen. *Semicond. Sci. Technol.* **8**, 824, 1993.

60 P. S. Wijewarnasuriya, M. D. Lange, S. Sivanathan, and J. P. Faurie. *J. Appl. Phys.* **75**, 1005, 1994.

61 P. E. Petersen. *J. Appl. Phys.* **41**, 3465, 1970.

62 P. R. Bratt. *J. Vac. Sci. Technol. A.* **1**, 1687, 1983; P. R. Bratt and T. N. Casselman. *J. Vac. Sci. Technol. A.* **3**, 239, 1985; K. Kosai and W. A. Radford. *J. Vac. Sci. Technol. A.* **8**, 1254, 1990.

63 L. O. Bubulac and C. R. Viswanathan. *Appl. Phys. Lett.* **60**, 222, 1992.

64 J. H. Tregilgas. *J. Vac. Sci. Technol.* **21**, 208, 1982.

65 J. Bajaj, W. E. Tennant, R. Zucca, and S. J. C. Irvine. *Semicond. Sci. & Technol.* **8**, 872, 1993.

66 D. R. Rhiger. *J. Electron. Mater.* **22**, 887, 1993.

67 J. A. Silberman, D. Laser, I. Lindau, W. E. Spicer, and J. A. Wilson. *J. Vac. Sci. Technol. A.* **3**, 222, 1985.

68 P. Morgen, J. Silberman, I. Lindau, W. E. Spicer, and J. A. Wilson. *J. Crystal Growth.* **56**, 493, 1982.

Deep Level Transient Spectroscopy: A Case Study on GaAs

WERNER K. GÖTZ and NOBLE M. JOHNSON

Contents

6.1 Introduction

Defects which introduce localized electronic states deep in the bandgap of semiconductors play an important role in the performance of devices. Both impurities and native defects—either intentionally or unintentionally introduced—can be sources of deep level defects, which are commonly referred to as electron or hole traps. The controlled introduction of deep level defects can be used to deliberately adjust the electronic or photoelectric properties of a semiconductor, examples of which are the growth of bulk semi-insulating GaAs for use as substrates in epitaxial growth, or the tuning of light emission or absorption in light-emitting devices or photodetectors, respectively. However, the presence of deep level defects can also degrade device performance; for example, deep level defects situated in the space charge layer of junction devices can act as carrier generation centers and thus increase leakage currents. Deep level defects also act as nonradiative recombination centers, which reduce the quantum efficiency in electro-optical devices.

In both cases, the characterization of deep level defects is essential. Deep level transient spectroscopy (DLTS), introduced by Lang in 1974, has become the most commonly used technique to investigate deep levels in semiconductors. With automated data acquisition techniques, DLTS has become a convenient, highly sensitive spectroscopic tool for deep level studies, with both individual components and complete DLTS systems commercially available.

This chapter reviews the application of DLTS for characterizing electronic defects in semiconductors, with emphasis on device fabrication, experimental setup, and measurement procedure. The technique is demonstrated by examining the results for an extensively studied deep level defect in GaAs: the EL2 center. The Bibliography lists several primary sources and review articles on DLTS which provide additional information and references to the substantial primary literature on this popular semiconductor spectroscopy.

6.2 DLTS Technique: General Features

With DLTS, the electronic properties of a deep level defect are determined by monitoring the transient response of an electrical signal as the electron occupancy of the defects relaxes from a spatial distribution of metastably charged centers by the thermal emission of charge carriers. As a purely operational or practical classification, defects with "deep levels" have energy levels in the semiconductor band gap typically greater than ~0.1 eV from either band edge; free carriers interact with these defects through capture and emission processes. A metastable occupancy is established by momentarily imposing conditions that favor the capture of free carriers. An electrical signal sensitive to these charge-state changes is readily obtained from semiconductor devices which possess a voltage-modulable space charge layer, such as a Schottky barrier diode or a p–n junction diode. The most commonly used electrical signal is the high-frequency differential capacitance of the device.

In a DLTS measurement, the test diode is initially under a reverse bias V_r, and deep levels in the space charge layer that are accessible to measurement are initially empty. The width W of the space charge region depends on the applied reverse bias voltage. For a short period of time, a pulse bias voltage V_p is applied which collapses the space charge layer and permits filling of some of these initially empty traps with majority carriers. Subsequently, the bias is returned to V_r, and the newly populated deep levels are in a metastable charge state. These defects relax to their originally empty state by charge emission.

Thermal emission of the metastable charge into the appropriate energy band and its rapid drift out of the space charge layer changes the total amount of charge in this layer. With DLTS, this charge change is usually monitored as a time-resolved change in the high-frequency (e.g., 1 MHz) differential capacitance of the space charge layer. Diodes for DLTS generally contain a sufficient density of fully ionized shallow dopants (e.g., donor concentration N_D) to guarantee that the capacitance change due to deep level carrier emission is small compared to the device capacitance. Then the transient response corresponds to an essentially constant number of deep levels, and the capacitance changes exponentially in time with relaxation of the trapped charge. The capacitance transient contains information on the density of traps (N_T) and on the activation energy for carrier emission. Thus, at low defect concentrations (i.e., $N_T << N_D$), the capacitance transient $C(t)$ displays the functional form

$$C(t) = C_0 - \delta C \exp(-e_n t) \tag{6.1}$$

where C_0 represents the steady-state high-frequency capacitance under reverse bias, δC is the total change of the capacitance due to deep level carrier emission, and e_n is the electron emission rate. From the principle of detailed balance, the thermal emission rate for electrons is given by

$$e_n = (\sigma_n v_n N_C/g) \exp\left[-(E_C - E_T)/kT\right] \tag{6.2}$$

where σ_n is the cross section for the capture of free electrons, v_n is the mean thermal velocity of electrons, N_C is the effective density of states in the conduction band, g is the (generally unknown) degeneracy factor for the defect, E_C is the energy position of the conduction band minimum, E_T is the energy level of the defect, k is Boltzmann's constant, and T is the absolute temperature. A similar expression applies for hole emission.

6.3 Fabrication and Qualification of Schottky Diodes

The GaAs material used in this study was grown on a degenerately doped n-type GaAs substrate by metalorganic chemical-vapor deposition. The growth temperature was 760 °C, and arsine and trimethylgallium (ratio 75:1) were used as source gases. A 4 μm thick epi-layer of GaAs was deliberately doped with Se, a shallow donor which provided the background n-type conductivity.

Schottky diodes were fabricated on the GaAs epilayers. Schottky diodes should be equipped with a low-resistance ohmic contact. To help achieve this, a degenerately doped substrate was used for the growth of the device. After cleaning the sample with solvents (acetone and methanol) and dilute HF, a 200 nm thick Au/Ge alloy layer (12 atom % Ge) was vacuum evaporated onto the back surface of the substrate, and the back contact was alloyed at 350 °C for 2 min in forming gas (85% N_2 and 15% H_2).

Schottky diodes were fabricated by evaporating an appropriate metal onto the n-type GaAs epi-layer to obtain a barrier for majority carriers at the contact interface. Prior to evaporation, the sample surface was cleaned with acetone, methanol, HF (to remove any oxide), and again with methanol. The metal was evaporated through a shadow mask with ~1 mm diameter holes to define a contact area. We evaporated a 50 nm thick Ti layer on the GaAs epi-layer and covered the Ti with 200 nm of gold to prevent oxidation of the Ti. The barrier height of this Schottky contact was equal to 0.72 ± 0.02 eV.

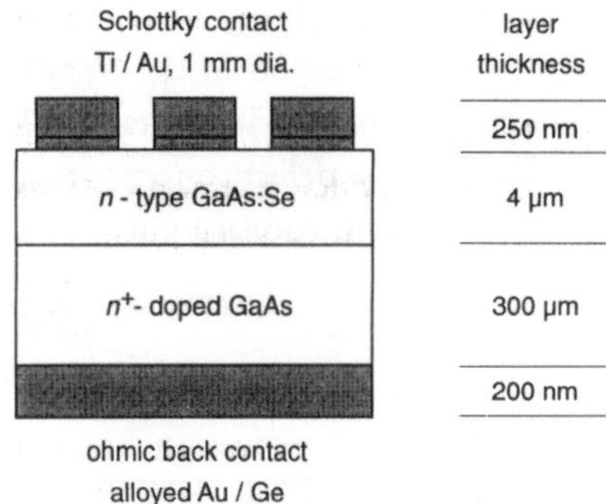

Figure 6.1   Schematic cross section of a Schottky diode on *n*-type GaAs.

For DLTS, the diode contacts deserve particular attention. The Schottky contact must be designed for low leakage current under large reverse biases. The contact resistance of the ohmic backside contact should be as close to zero as possible to avoid any additional series resistance which could complicate the interpretation of the 1 MHz capacitance measurement. Besides alloying, high-dose ion implantation of an appropriate dopant to create an n^+ or p^+ layer (with furnace activating) can be used to form an ohmic back contact.

A cross section of the Schottky diode used in this study is shown in Figure 6.1. The performance of the diode was evaluated with standard current-voltage (I–V) and capacitance–voltage (C–V) techniques. The diode exhibited excellent current rectification with low series resistance ($\sim$11 Ω) at large forward biases and low leakage current (e.g., $\leq 3 \times 10^{-7}$ A/cm^2 for a reverse bias of 10 V). (See Figure 6.2*a*.)

Whereas standard I–V measurements characterize essentially the dc performance of the Schottky diode, the C–V analysis serves to check its high frequency performance. In a capacitance measurement (C–V or DLTS), the differential capacitance, $C = dQ/dV$, is monitored. The current response to a small ac signal (test level $\leq$ 100 mV) is measured, which, due to the impedance of the device, contains both capacitive and resistive components. Since the DLTS measurement is conducted with a 1 MHz voltage test signal, the same frequency (1 MHz, 17 mV rms) was selected for the C–V analysis. The diodes display typical C–V characteristics for a uniformly doped semiconductor (Figure 6.2*b*). The phase angle of the measured ac impedance under a reverse bias of 8 V was –88.6°, which indicates that the impedance is dominated by the capacitive component.

The DLTS measurement requires that time constants associated with normal diode performance, such as the response of the space charge depth to the 1 MHz test signal, be short compared to time constants associated with deep level carrier emission. Therefore, a careful analysis of the I–V and C–V characteristics of the test device is obligatory. High series resistance of the substrate layer or high contact resistance of the backside contact can easily lead to capacitance transients that can

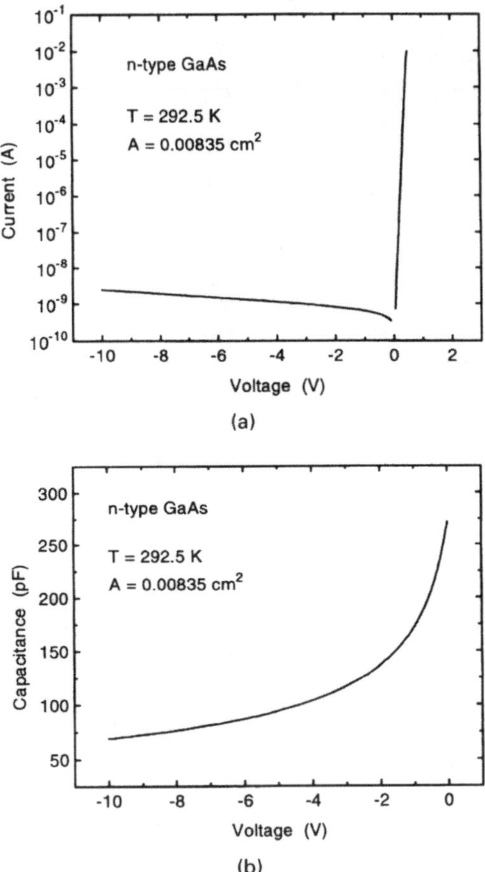

Figure 6.2   (*a*) Current-voltage and (*b*) capacitance–voltage data for a Schottky diode on *n*-type GaAs.

be misinterpreted, in the extreme leading to the apparent characterization of non-existent deep level defects.

The C–V data were analyzed to determine the depth distribution of the net doping concentration. The concentration profile is displayed in Figure 6.3 and reveals an essentially uniform donor concentration of 8.8×10^{15} cm^{-3} over the measured depth interval from 0.4 to 1.3 μm.

The depletion depth is a function of the applied reverse bias, $W(V)$ and is a key parameter in selecting the spatial interval over which deep levels can be detected in a DLTS measurement. For the C–V analysis, the high frequency (HF) capacitance was modeled as the differential capacitance of a Schottky diode, which is given within the depletion approximation as

$$C_{HF} = \frac{\varepsilon_s A}{W} \tag{6.3}$$

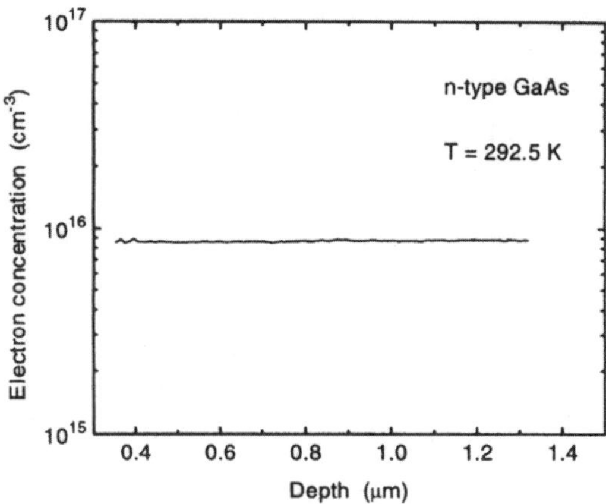

Figure 6.3 Depth profile of the net donor concentration in a Schottky diode on *n*-type GaAs.

where ε_S is the permittivity for the semiconductor (e.g., $\varepsilon_S = 13.1 \cdot \varepsilon_0$ for GaAs, where ε_0 is the permittivity of free space), A is the area of the Schottky contact, and W is the width of the depletion layer. For the present study, the spatial interval for the DLTS measurement was determined by a reverse bias voltage $V_r = -8$ V ($W \approx 1.2$ µm) and a pulse bias voltage $V_p = -2$ V ($W \approx 0.7$ µm).

6.4 DLTS System

The essential components of a DLTS measurement system are shown in Figure 6.4. The test diode is mounted on a sample holder in a variable temperature cryostat. Accurate monitoring of the diode temperature (e.g., resolution of 0.1 K) is essential, since the emission rate, Equation 6.2, depends exponentially on temperature. This can be accomplished by mounting a calibrated temperature sensor in close proximity to the diode or in a position thermally equivalent to that of the diode. Relying on the temperature reading of the cryostat controller may not be satisfactory since the controller sensor is usually located at some intermediate position between the holder and the heating elements and, therefore, may not accurately measure the diode temperature, particularly during a temperature ramp. The temperature range for a DLTS measurement should be selected to be as wide as possible in order to access the widest energy interval in the semiconductor bandgap over which to detect deep levels. The low-temperature limit is determined by the onset of freeze-out of the shallow dopants, if not by the low-temperature limit of the cryostat; and the high-temperature limit may result from the degradation of diode performance (e.g., excess leakage current), if not by the high-temperature limit of the heater stage.

At the heart of the DLTS system is a capacitance meter or bridge, which combines

     DEEP LEVEL TRANSIENT SPECTROSCOPY . . .   Chapter 6

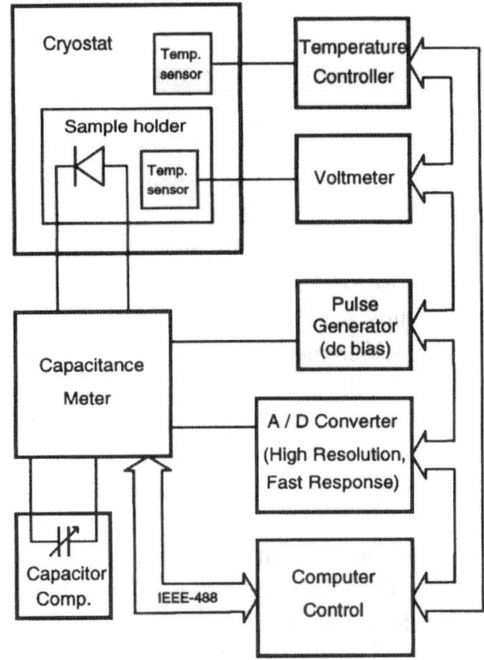

Figure 6.4   Basic components of a DLTS system.

high sensitivity with a fast response to changes in diode capacitance, and fast recovery from overload conditions. Commercially available capacitance meters also permit application of the diode bias, both the reverse and pulse biases, through the instrument circuitry. Enhancement of the sensitivity is achieved in such instruments through a differential input, which permits the attachment of an external variable capacitor to compensate the diode capacitance at a given reverse bias; the compensation permits the selection of a higher sensitivity range. The time-varying voltage applied to the diode is typically supplied by a pulse generator with a variable dc baseline voltage.

The capacitance transients can be transmitted via the fast analog output of the capacitance meter to a high-resolution A/D converter (e.g., a digital oscilloscope), which should be capable of averaging many capacitance transients before transmitting the averaged transient waveform to a computer (e.g., via an IEEE-488 bus). The timing of the transients is controlled by the pulse generator, which triggers the fast A/D converter for recording a capacitance transient synchronized to the trailing edge of the pulse bias. The computer can be interfaced with all instruments of the DLTS system to accomplish completely automated measurement control and data acquisition. The averaged capacitance transient waveforms are recorded over a range of diode temperatures and stored for subsequent data analysis.

6.5 DLTS Measurement Procedure

In DLTS, the (averaged) capacitance transient is recorded over a range of temperatures. The common practice is to slowly vary the diode temperature over the selected range such that any given transient is recorded over a very narrow temperature interval (e.g., <1 K). An alternative approach, used in the present study, is to step the temperature from setpoint to setpoint and record the capacitance transient at a series of constant temperatures. At each set point of the temperature controller, the diode temperature is allowed to stabilize before initiating the averaging of individual capacitance transients. In the present study, we selected a temperature range from 100 K to 450 K, and the temperature was incremented in 1 K steps. Prior to the DLTS measurement, the diode characteristics were evaluated (Section 6.3) at the temperature limits in order to quality the device for further study.

For the DLTS measurement, the reverse bias voltage V_r was set at –8 V and the pulse bias voltage V_p was set to –2 V. The voltage sequence and timing of the measurement are illustrated in Figure 6.5.

The trap-filling voltage pulse V_p was applied for 1 ms, and then the voltage was switched back to the reverse bias voltage V_r. The duration of the filling pulse t_p should be set long enough to guarantee complete trap filling during the pulse. We tested this by observing a saturation for the capacitance transient with respect to t_p. The pulse period was 100 ms. At the trailing edge of the filling pulse, the A/D converter was triggered to initiate the recording of the capacitance transient. Approximately 400 transients were averaged at each temperature. Then the averaged transient was transmitted to the computer and stored. For the A/D conversion, we chose a time interval of 40 ms on each transient, with 50 µs between points, which were scaled with the capacitance meter range calibration to obtain a stored set of capacitances versus time at each temperature.

An example of an averaged transient is displayed in Figure 6.6. The individual transients which contributed to the averaged waveform were collected at a diode temperature of 385.2 K. In the upper part of Figure 6.6, we show the voltage sequence applied to the test device.

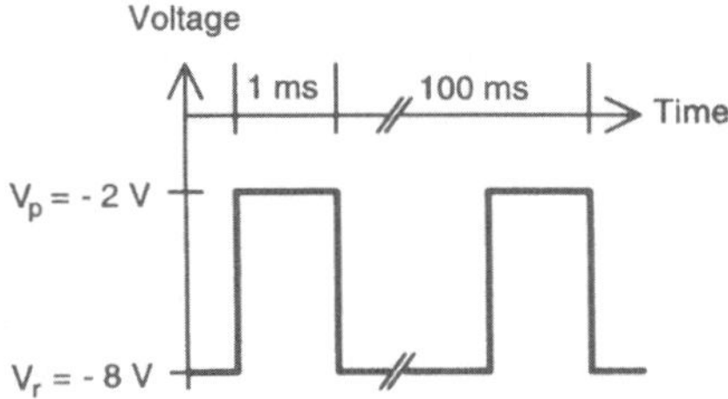

Figure 6.5   **Voltage sequence applied to an *n*-type GaAs Schottky diode for DLTS measurement.**

6.6 Data Analysis

DLTS Spectrum

The most widely adopted way to display DLTS data is with a DLTS spectrum. For this purpose, a DLTS signal is first obtained by correlating the capacitance transient with a simple weighting function, which consists of delayed delta functions

$$w(t) = \delta(t - t_2) - \delta(t - t_1) \tag{6.4}$$

where t_1 and t_2 are delay times after the filling pulse. The position of the delay times is indicated in Figure 6.6. This weighting function $w(t)$ can be readily implemented in analog instrumentation by monitoring the capacitance transients with gated integrators (e.g., a double boxcar averager), or digitally on a computer from the averaged transient waveform. In a DLTS measurement, signal processing is key for extracting useful trap information from a weak, noisy transient signal.

The convolution of the capacitance transient (Equation 6.1) with the weighting function (Equation 6.4) over the time interval $0 < t_1 < t_2 < \tau$ produces the DLTS signal

$$\Delta C \equiv C(t_2) - C(t_1) = \delta C\left[\exp(-e_n t_1) - \exp(-e_n t_2)\right] \tag{6.5}$$

The key feature of the DLTS signal ΔC is that it goes through a maximum at that temperature for which the trap emission rate, e_n, equals the "emission rate window," e_0, which is set by the delay times

$$e_0 = \frac{\ln(t_2/t_1)}{t_2 - t_1} \tag{6.6}$$

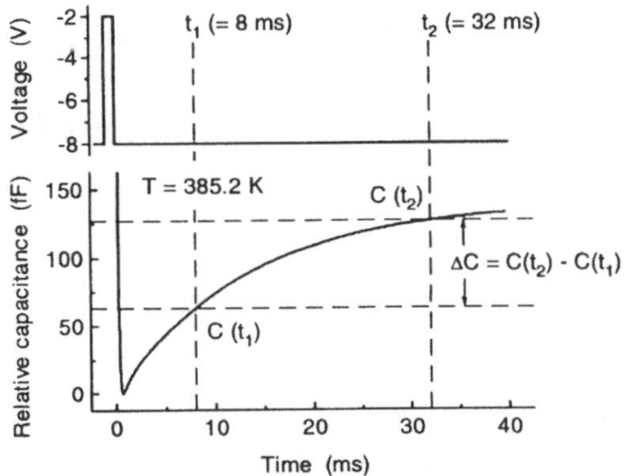

Figure 6.6   Voltage sequence (*top curve*) and averaged capacitance transient (*bottom curve*) versus time, with the time reference $t = 0$ at the trailing edge of V_p. The dashed lines indicate the delay times t_1 and V_2 and the corresponding capacitances $C(t_1)$ and $C(t_2)$ used to generate the DLTS signal ΔC.

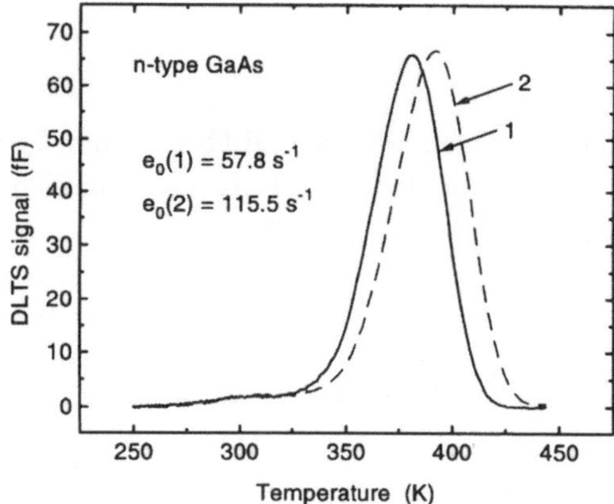

Figure 6.7   DLTS spectra for a Schottky diode on *n*-type GaAs. The spectra are for two different emission rate windows: $e_0(1)$ (t_1 = 8 ms, t_2 = 32 ms) and $e_0(2)$ (t_1 = 4 ms, t_2 = 16 ms).

Thus, a discrete defect level introduces a peak in a DLTS spectrum. This is illustrated in Figure 6.7 for the Schottky diode on *n*-type GaAs. The spectrum displays two peaks: a strong, well-resolved peak near 390 K and a very weak unresolved peak near 300 K. The unresolved peak is not considered in this study.

The large emission peak has been extensively studied in GaAs and is labeled the EL2 center (see Section 6.7). The peak position is 381 K ($T_{peak}(1)$) for the emission rate window $e_0(1)$ and 392 K ($T_{peak}(2)$) for $e_0(2)$. Since $e_0(1) < e_0(2)$, the peak in the DLTS signal for the EL2 center appears in the DLTS spectrum for $e_0(1)$ at a lower temperature than for $e_0(2)$ ($T_{peak}(1) < T_{peak}(2)$).

Activation Energy for Thermal Emission

The activation energy for thermal emission from a deep level defect is commonly obtained from an Arrhenius analysis of the emission probability. However, the temperature dependence of the capture cross section, $\sigma_n(T)$, must be either assumed or separately determined.

The temperature dependence of the capture cross section for some capture mechanisms may be described as

$$\sigma_n(T) = \sigma_{n,0} T^{\alpha} \tag{6.7}$$

where $\sigma_{n,0}$ is a constant and the parameter α accounts for the temperature dependence. Electron capture involving excited electronic states of the defect (cascade process) is an example of the above dependence on temperature. For the cascade capture process, α ranges from -2 to -3.

Multiphonon capture processes can give rise to a thermally activated cross section, which may be described as

$$\sigma_n(T) = \sigma_{n,0} \exp(-E_\sigma/kT) \tag{6.8}$$

where E_σ is some energy barrier to carrier capture.

In most cases, the capture mechanism is unknown and a temperature independent mechanism is assumed for the Arrhenius analysis. This approach was adopted in the present study.

For the Arrhenius analysis we use Equation 6.2. Thermodynamically, the trap energy is identified with the Gibbs free energy of ionization, ΔG_n,

$$\Delta G_n = E_C - E_T = \Delta H_n - T\Delta S_n \tag{6.9}$$

where ΔH_n and ΔS_n are the changes in enthalpy and entropy, respectively. Combining Equations 6.2, 6.7, and 6.9 and using the temperature dependencies of the prefactors ($v_n = vT^{1/2}$, $N_C = NT^{3/2}$) lead to the following expression for the emission rate (the degeneracy factor is assumed to be unity):

$$e_n = \sigma_{n,0}vNT^{2+\alpha}\exp(\Delta S_n/k)\exp(-\Delta H/kT) \tag{6.10}$$

where v and N are known, temperature-independent constants. Equation 6.10 reveals that the Arrhenius analysis does not directly yield ΔG_n. However, ΔS_n is generally assumed to be negligible, so that $\Delta H_n \approx \Delta G_n \equiv \Delta E_T$.

With the assumption of a temperature-independent cross section ($\alpha = 0$), Equation 6.10 can be expressed as

$$\ln(e_n T^{-2}) = \ln(\sigma_{n,0}vN) - \frac{\Delta E_T}{k}\cdot\frac{1}{T} \tag{6.11}$$

Changing the emission rate window e_0 shifts the temperature of the peak in the DLTS spectrum, as illustrated in Figure 6.7. Since at the peak the emission rate $e_n(T)$ equals the rate window e_0, recording DLTS spectra for a set of selected rate windows yields the temperature dependence of the emission probability. An alternative approach is to fit each averaged capacitance transient with a single exponential function to obtain the decay time. This technique was employed in the present study. A least-squares fit of an exponential function was performed on a representative subset of averaged capacitance transients in the temperature range from 360 K to 400 K, where electron emission from the EL2 center is dominant. The results are shown in Figure 6.8 as an Arrhenius plot of $\ln(e_n T^{-2})$ versus $1/T$. The experimental data fit a straight line (obtained from a linear regression analysis), the slope of which yields the activation energy for electron emission from the EL2 center, 0.76 ± 0.01 eV. The thermal activation energy is the most commonly quoted parameter to characterize a deep level defect by DLTS.

The capture cross section can be estimated by extrapolation of the straight line to the intercept with the abscissa ($x = 0$), which yields $(9.2 \pm 0.2) \times 10^{-14}$ cm^2.

The uncertainties given for ΔE_T and σ_n refer to the measurement uncertainty, but do not account for the uncertainties that originate from the unknown temperature dependence of the capture cross section and neglect of ΔS.

In addition, emission rates determined with DLTS are measured under the high electric field present in a space charge region of a semiconductor. For uniform

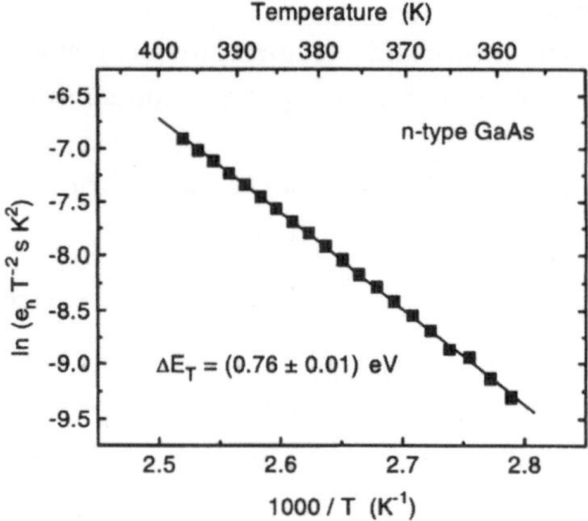

Figure 6.8   Arrhenius analysis of the electron emission rate e_n for the peak in the DLTS spectrum of Figure 6.7. The symbols refer to experimental data, while the solid line represents a linear regression fit to the data. The slope of the straight line yields the activation energy for electron emission from the trap.

doping, the magnitude of the electric field in the space charge region varies approximately linearly with depth from zero at the depletion edge ($x = W$) and is maximum at the interface (e.g., ~10^6 V·cm^{-1}). Some traps experience a Coulomb attraction to majority carriers during carrier emission, which affects the emission rate; this process is called the Poole–Frenkel effect.

Trap Densities

Densities of deep level defects can be determined with good accuracy. The δC in Equation 6.5 is the following expression (for uniform N_D and $N_T << N_D$):

$$\delta C = \frac{N_T C_0^3}{2N_D \varepsilon_S^2 A^2}(x_r^2 - x_p^2) \tag{6.12}$$

The parameters x_r and x_p are called the "crossover points" in the space charge region of the Schottky diode and mark the intersection of the quasi Fermi level with the defect energy level. The depth x_r marks the intersection under reverse bias (V_r), and x_p marks the intersection under pulse bias voltage (V_p). These crossover points define the "spatial observation window," in which deep levels are probed by the DLTS measurement. The magnitude of the DLTS signal at an emission peak in a DLTS spectrum is related to the trap density N_T by the following expression (Equations 6.5 and 6.12):

$$N_T = \frac{2N_D \varepsilon_S^2 A^2 \Delta C(T_{\text{peak}})}{C_0^3}(x_r^2 - x_p^2)^{-1}\left[\exp(-e_0 t_1) - \exp(-e_0 t_2)\right]^{-1} \tag{6.13}$$

where C_0 is the steady-state device capacitance under reverse bias at the peak temperature. For uniform doping, the crossover points x_r and x_p can be calculated from

$$x_{r/p} = W(V_{r/p}) - \sqrt{\frac{2\varepsilon_S(\Delta E_{FS} - \Delta E_T)}{q^2 N_D}} \qquad (6.14)$$

where ΔE_{FS} is the quasi Fermi level in the semiconductor, which depends on N_D and T. From Equations 6.13 and 6.14, the density of the EL2 center which gives rise to the DLTS signal in Figure 6.7 is calculated to be $N_T = (8.5 \pm 0.5) \times 10^{13}$ cm^{-3}.

6.7 EL2 Center

The EL2 center is the dominant deep donor defect in high-resistivity bulk GaAs. To produce semi-insulating bulk material, it is deliberately introduced by growing under arsenic-rich conditions to compensate residual acceptors. It has been shown that the EL2 defect is a native lattice defect not related to impurities. Although the EL2 center has attracted much attention among researchers for more than 30 years, its microscopic structure is not yet fully understood. However, two models for the EL2 centers have been proposed which seem to be consistent with the available experimental data: (1) a point defect consisting of an isolated As atom located on a Ga lattice site (As_{Ga}) or (2) a point defect complex consisting of an As_{Ga} antisite correlated with an As interstitial (As_{Ga}–As_i). Besides the normal (fundamental) configuration, a metastable state is associated with this defect. Electrical and optical properties of the fundamental and metastable configurations are significantly different.

The fundamental EL2 configuration is a double donor introducing two deep localized levels into the bandgap of GaAs. The (0/+) level of the EL2 center (E_C – 0.76 eV) was used in the present study to demonstrate the DLTS technique. The second level (1/21) lies about 0.25 eV below the first level. Besides DLTS, optical techniques and especially EPR have been fruitfully employed to investigate the intriguing physical properties of the EL2 center in GaAs and to establish microscopic structural models for this defect.

6.8 Summary

DLTS is used to characterize the electronic properties of deep level defects in semiconductors. The measurement is performed on semiconductor devices that possess a voltage-modulable space charge layer. Deep level defects are detected by the relaxation of device capacitance in response to the thermal emission of trapped charge. The measurement furnishes phenomenological parameters which relate to the location of a deep level defect in the semiconductor band gap and to its interaction with free carriers. These parameters include ionization energies, cross sections for carrier capture, and depth distribution (a property of the material rather than of the defect). The characterization of the EL2 center in GaAs served as a demonstration of the capabilities of the DLTS technique.

Acknowledgment

The authors are pleased to acknowledge partial support from AFOSR (Contract No. F49620-91-C-0082). W. Götz is also pleased to acknowledge support from the Alexander von Humboldt Foundation in Germany.

Bibliography

DLTS

Johnson, N, M. "Measurement of Semiconductor-Insulator Interface States by Constant-Capacitance Deep Level Transient Spectroscopy," *Journal Vacuum Science and Technology.* **21** (2), 303–314, 1982. A review of the DLTS measurement and analysis for the determination of interface state distributions in metal–insulator–semiconductor structures.

Johnson, N. M., D. J. Bartelink, R. B. Gold, and J. F. Gibbons. "Constant-Capacitance DLTS Measurement of Defect Profiles in Semiconductors," *Journal of Applied Physics.* **50** (7), 4828–4833, 1979. Describes the constant capacitance mode of the DLTS measurement and its formulation for determining depth profiles of deep levels.

Lang, D. V. "Fast Capacitance Transient Apparatus: Application to ZnO and O Centers in GaP *p–n* Junctions," *Journal of Applied Physics.* **45** (7), 3014–3022, 1974; "Deep Level Transient Spectroscopy: A New Method to Characterize Traps in Semiconductors," *Journal of Applied Physics.* **45** (7), 3023–3032, 1974. The original articles introducing the DLTS technique.

Lefevre, H., and M. Schulz. "Double Correlation Technique (DDLTS) for the Analysis of Deep Level Profiles in Semiconductors," *Applied Physics.* **12** (1), 45–53, 1977. Introduces the double-correlation DLTS technique for determining the depth distribution of bulk deep levels in semiconductors.

Miller, G. L., V. Lang, and L. C. Kimerling. "Capacitance Transient Spectroscopy," *Annual Review of Material Science.* 7, 377–448, 1977. An early review of the DLTS technique and its application to bulk defect characterization.

Pensl, G. "Measurement Methods." In *Landoldt-Börnstein: Data Tables on Impurities and Defects*, Series III, Vol. 22, Part b. (O. Madelung and M. Schulz, Eds.) Springer-Verlag, Berlin/Heidelberg, 1989, chapt. 3. A recent comprehensive compilation of electrical measurements on semiconductors, including many variations of the DLTS technique, with an extensive list of references.

Williams, R. "Determination of Deep Centers in Conducting Gallium Arsenide," *Journal of Applied Physics.* **37** (9), 3411–3416, 1966. Historically interesting as the first experimental demonstration of the detection and characterization of a deep level defect with transient capacitance measurement.

EL2 Center

Baraff, G. A. "The Mid-Gap Donor Level EL2 in GaAs: Recent Developments," in *Deep Centers in Semiconductors*, 2nd ed. (S. T. Pantelides, Ed.) Gordon and Breach Science Publishers, New York, 1992, pp. 547–589. A continuation of the article by G. M. Martin and S. Makram-Ebeid. It reviews the more recent observations and discusses the microscopic structure for the EL2. In addition, a comprehensive bibliography on the topic is included.

Martin, G. M., and S. Makram-Ebeid. "The Mid-Gap Donor Level EL2 in GaAs," in *Deep Centers in Semiconductors*. (S. T. Pantelides, Ed.) Gordon and Breach Science Publishers, New York, 1986, pp. 399–487. A comprehensive review of the electrical and opto-electronic properties of the EL2 center up to 1983.

Appendix: Technique Summaries

The technique summaries in the following pages of this appendix include many one page summaries marked with an asterisk. These techniques are fully described in the *Encyclopedia of Materials Characterization* by C. Richard Brundle, Charles A. Evans, Jr., and Shaun Wilson. The remaining summaries are for techniques which do not appear in that volume.

Auger Electron Spectroscopy (AES)* 1

In Auger Electron Spectroscopy (AES), a focused beam of electrons (typically a few keV to a few 10's of keV in energy) strikes the sample, causing electrons to be ejected. Some of these (the Auger electrons, named after Pierre Auger who first observed them) have kinetic energies characteristic of the atoms from which they came and so identification of the elements present in the sample can be made directly from the measurement of these energies. On a finer scale, it is sometimes also possible to determine the chemical state of the element from small shifts in the Auger energies caused by the chemical bonding of the element (chemical shift). For a solid, AES probes 2–20 atomic layers deep, depending primarily on the kinetic energy of the ejected Auger electron concerned, and somewhat on the material. With appropriate standards the relative concentration of the elements present within the probing depth can be estimated from the relative intensities of the Auger peaks in the spectrum. The great strength of AES is the high spatial resolution achievable through use of a focused beam (down to 10's of nm), and the ability to combine measurement with ion sputtering material removal, to obtain a three-dimensional elemental profile. The main use is with metals and conducting or semiconducting inorganic materials, since beam damage and charging can be issues with non-conducting and organic materials. Many Auger spectrometers are designed to maximize sensitivity (typically down to 100 ppm of element concentration) and speed of analysis, and do not have the spectral resolution to use chemical shifts for chemical state identification. Auger is capable of identifying all elements, except hydrogen and helium.

PHILLIP NIEDERMANN

Ballistic electron emission microscopy (BEEM) is a technique used to investigate the electronic structure of buried interfaces with high spatial resolution. Primarily, BEEM can be used to locally measure and produce images of the height of subsurface electronic barriers, such as Schottky barriers on n- or p-doped semiconductors. In addition to that, interface electronic state density, hot carrier transport, and doping profiles can be investigated by BEEM. The basic measuring instrument is a scanning tunneling microscope (STM) used in a three-terminal configuration with the tip as electron emitter, a base through which electrons are transmitted without scattering (hence the term ballistic), and a collector separated from the base by an electronic barrier (the specimen), such as a Schottky barrier. The spatial resolution depends on the carrier scattering properties of the base; it can be as high as ~1 nm. Since the BEEM image of an interface is taken while the STM tip is scanning the surface, an atomic-resolution topographic image of the surface can be obtained simultaneously with the interface image.

Information	Spatially resolved barrier height for Schottky barriers, heterojunctions, p–n junctions, etc. Interface electronic state density. Hot carrier scattering. Recombination/generation currents
Destructive	No
Environment	Vacuum, inert atmosphere or air
Sample requirements	Schottky contact or other electronic barrier under a conducting layer thinner than a few tens of nm. Zero-voltage impedance across the barrier >100 kΩ (for low barriers, small device area and/or low temperature required)
Lateral resolution	Down to 1 nm
Imaging	Inherent in technique
Measuring equipment	STM
Cost	As for an STM, $50,000–$50,000

Introduction

For surface analysis, a broad range of well-established techniques exist for the study of electronic and structural properties of surfaces and thin overlayers. Interfaces between bulk materials, though important for electronic and other material properties, are much more difficult to access with microscopic or spectroscopic methods. The measurement of electronic barrier heights with submicrometer resolution was not possible before BEEM was invented.

As with STM, BEEM[1] can be performed in air, ultra-high vacuum, or liquids at temperatures ranging from cryogenic to above room temperature. In general terms, BEEM is a method to inject hot carriers with controllable energy into a solid and to measure their collection efficiency directly or via a secondary process. There is an entire class of BEEM-type experiments, including ballistic hole emission microscopy, interface band structure spectroscopy, and electron-electron scattering spectroscopy. Since it is a very recent development, with much exploratory research currently underway, BEEM will be further expanded to include even more variations and uses in the future.

Experimental Setup

The experimental setup of BEEM is shown in Figure 1 for a Schottky barrier on an *n*-doped semiconductor. A constant current passes, via tunneling, from the tip of an STM into the sample, regulated by the feedback circuitry as in a standard STM imaging mode. The sample is a layered system having at least two layers separated by an electronic barrier. Until now, most measurements have been done on *n*-type or *p*-type Schottky barriers, with the metal layer ("base") grounded and the semiconductor ("collector") held at virtual ground and connected to an electrometer amplifier. Electrons that tunnel into the base may travel without energy loss to the metal-semiconductor interface. If their energy is high enough, they may enter the semiconductor and drift away from the interface under the influence of the electric field in the depletion zone of the semiconductor. A current-voltage converter detects the collector current, which is typically on the order of a few picoamperes and can be measured either as a function of tunnel voltage so as to

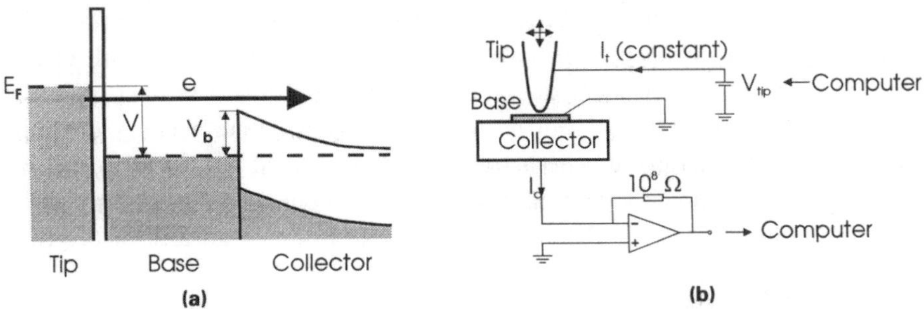

Figure 1   (*a*) **Energy level diagram for BEEM process, (*b*) Schematic of experimental setup.**

obtain BEEM spectra or as a function of position at a constant tunnel voltage so as to obtain interface images.

Physical Principle

The transmission of an electron can be described in three steps:

1 Tunneling from tip to base. This produces a density of hot electrons in the base that decreases exponentially below the Fermi energy of the tip and away from the surface normal.

2 Transport through the base. If the base is thin, as compared to the elastic mean free path, a large fraction of the electrons can reach the interface without scattering. In gold, for example, the mean free path is about 130 Å,[2] and a base layer of 100 Å is adequate for a typical BEEM measurement with a tunnel current on the order of 1 nA.

3 Transmission across the interface. For an ideal interface, transmission can occur if an electronic state in the semiconductor is available with the same energy and the same **k** vector component parallel to the interface ($k_{\parallel}$) as the impinging electron. If the kinetic energy of the electron is just barely above the conduction band minimum of the semiconductor, $k_{\parallel}$ conservation dictates that only those electrons whose **k** vector lies in a narrow cone in the base can be transmitted.

A model for the collector current in BEEM has been given by Bell and Kaiser (BK model).[3] In this model, the tunneling between tip and base is described by the transmission probability $D(E_x)$ as a function of tip normal kinetic energy E_x, assuming a planar geometry.

The onset of the collector current, as predicted from this model, is quadratic, in agreement with experiment. The local barrier height can therefore be determined to a good approximation by plotting dI_c/dV versus V (see Figure 1) and performing a linear fit in the onset region.

Among the physical effects that should be taken into account in addition to those in this model are elastic scattering in the base[4] and quantum mechanical reflection at the interface.[5]

Sample Requirements

Samples that can be studied are Schottky barriers or other barriers covered by a thin metallic layer of a thickness comparable or lower than the mean free path of electrons in the eV range—typically on the order of 100 Å. If the base layer is an inert material such as gold, BEEM can be performed in air, rather than in ultra-high vacuum or inert atmosphere. The barrier impedance should be 100 kΩ or higher, requiring Schottky diodes with a barrier lower than about 0.7 eV to be cooled to a low temperature (77 K) for BEEM measurements (assuming an area of 0.1 cm^2).

Spatial Resolution

The BK model predicts a very high spatial resolution for a semiconductor with a conduction band minimum at $k_{\parallel} = 0$. A small critical angle above threshold implies that interface changes on a scale much smaller than the thickness of the base are measurable, putting the spatial resolution of the technique in the nanometer range. This is an ideal case; in practice, the resolution may be only in the 10 nm range for a variety of possible reasons:

1 The conduction band minimum is not at $k_{\parallel} = 0$, as is the case for Si(111)

2 Multiple elastic scattering occurs in the base before an electron reaches the interface; in this case, the grain structure of the base may strongly influence the resolution

3 Electrons are scattered at the interface, relaxing the conservation of $k_{\parallel}$. This is expected, in particular, for a nonepitaxial interface.

The attainable resolution is thus dependent on the nature and quality of the base layer and is subject to ongoing research.

In Schottky barriers, variations of barrier height at the interface are screened in the semiconductor. A small spot with a reduced barrier may be "pinched off," with a saddlepoint occurring in the potential barrier.[6] The lower the electric field near the interface, the more pronounced this effect.

Related Techniques

As already mentioned, BEEM is a general-purpose technique to inject low-energy electron or hole beams into a layered solid system; a number of BEEM-related techniques have already been demonstrated. Schottky barrier heights on p-doped semiconductors have been successfully measured[7] at 77 K using ballistic hole spectroscopy, in which hot holes are injected into the base by positively biasing the tip, in direct analogy to electron BEEM.

If the tip is biased *positively* on an n-doped Schottky diode, a collector current can again be observed that has the same sign as the BEEM current. This "reverse BEEM" current is created by the decay of hot holes in the base via collisions with electrons[8]—in analogy to an Auger process—and provides a tool for the study of carrier scattering in the metal. A similar experiment has also been demonstrated on p-doped Schottky diodes.[8]

Applications

BEEM has already been used to characterize Schottky barrier height and its spatial homogeneity in a variety of metal/n-type semiconductor systems, such as Au/Si(100), on which the initial BEEM work was done.[1]

The PtSi/n-Si(100) system[9, 10] is a sharp, easy-to-form interface that has important applications because of its high Schottky barrier of ~0.90 eV. It has been studied by BEEM for a range of silicide thicknesses.[9] The silicide layer on these diodes was found to have a relatively complex morphology, containing granular regions of local epitaxy. This granularity is reflected in BEEM images (Figure 2). For this system, the spectral behavior depended on whether samples had been exposed to air or had been produced and studied entirely in situ. In the former case, the spectra showed considerable variations in threshold and intensity, probably due

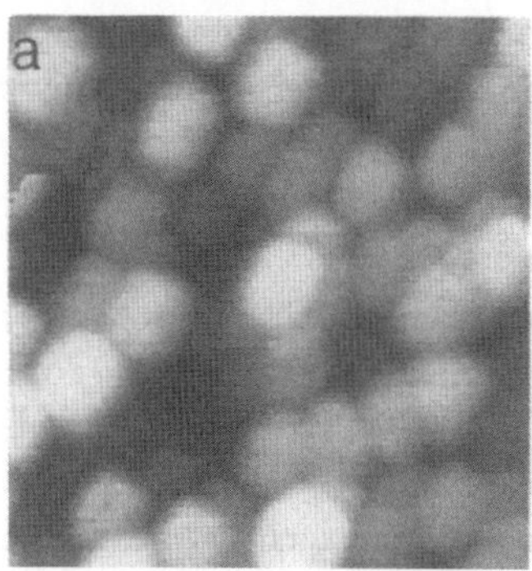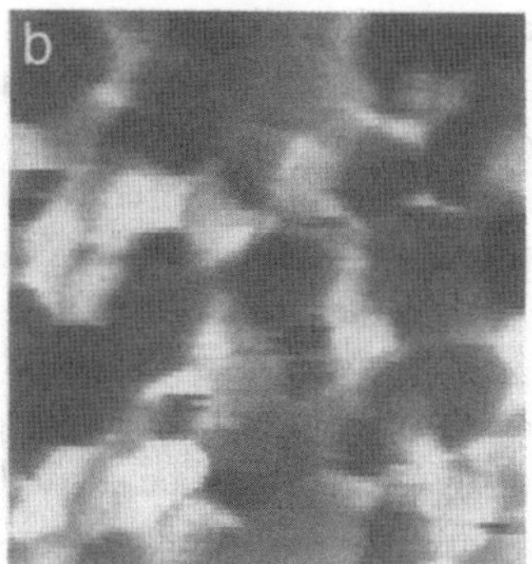

Figure 2 (*a*) STM topograph and (*b*) collector current image taken on a sample with 50 Å of PtSi on *n*-Si(100) showing strong variations of the BEEM current, which are correlated to the grain structure of the layer. Scales from black to white are (*a*) 40 Å and (*b*) 57 pA. Tunnel voltage and current were 1.5 V and 1 nA, respectively.

to surface impurities (Figure 3). On the other hand, for the in situ produced samples, the spectral onsets were constant as a function of position while the intensity still varied strongly (up to two orders of magnitude) over a scale as small as 100 Å. The variations in intensity are closely related to the granular topography and are probably due to different morphologies of the interfaces on differently oriented grains. The average collector current intensity was about an order of magnitude higher than for the ex situ samples of the same thickness, indicating a longer electron mean free path for the former.

Ion bombardment of silicon is known to introduce *n*-type defects. For a Schottky barrier, this is believed to create a high electric field near the interface, which in turn reduces the effective barrier height through the effect of the image force and electron tunneling. A BEEM study[10] on PtSi/*n*-Si(100) diodes subjected to a relatively strong rf plasma treatment showed the Schottky barrier to be lowered (Figure 3), in reasonable agreement with diode *I*–*V* measurements, indicating that the barrier lowering occurred uniformly. Modeling the BEEM spectra for a high-field barrier showed that a sizable fraction of the BEEM current was due to tunneling across the interfacial barrier. If the interface electric field is high, the "pinch-off" effect mentioned previously is strongly reduced, and intentionally introducing high doping should allow one to measure Schottky barrier variations on a very small scale in non-uniform systems, which are probably more common than is generally believed.

Besides Schottky barriers, other types of electronic barriers can be studied by BEEM. In particular, BEEM has been used to characterize *p*–*n* junctions in Si.[11] Application to thin insulating (e.g., oxide) barriers should also be possible.

Another fascinating development is the modification of interfaces using the STM itself. A study has been performed in which nanoscale features have been

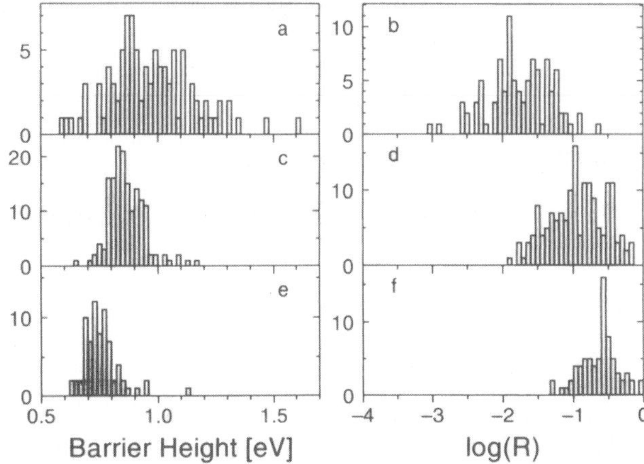

Figure 3 Histograms of barrier heights V_b and scale factors R for differently prepared PtSi/n-Si(100) Schottky diodes fabricated by Pt sputter-deposition on Si and annealing at 500 °C. In situ produced diodes (c,d) had a much more narrow V_b distribution and a higher R than ex situ diodes (a,b) (PtSi thickness 50 Å). Exposing the Si substrates to strong rf sputter discharge cleaning before producing PtSi diodes (30 Å) resulted in Schottky barrier lowering (e,f).

written and imaged by BEEM on an Au-Si(100) interface.[12] This opens the possibility of subsurface nanofabrication.

Conclusion

The ability of BEEM to locally measure Schottky barrier heights is now well demonstrated for a variety of materials, showing generally good agreement with other types of measurements. Beyond barrier height and interface electronic structure, BEEM can be used to investigate hot carrier transport in metal films, making this technique useful for studying film growth properties with nanometer spatial resolution.

The future will bring more uses of BEEM and related techniques as more is learned on hot carrier transport physics in metals and semiconductors, Schottky barrier local structure, and interface electronic structure. The study of fully epi-taxially grown systems will deepen our understanding of the ballistic transmission process. The realm of BEEM has already been expanding and will continue to do so, from Schottky barriers on homogeneously doped semiconductors to a variety of specially designed electron barrier systems. The combination of BEEM, TEM, surface analytical tools, and other experimental techniques will deepen our understanding of specific metal-semiconductor systems.

Acknowledgments

The author would like to acknowledge the efforts of his collaborators at the University of Geneva, Lidia Quattropani and Ø. Fischer, and to thank L. D. Bell,

M. H. Hecht, W. J. Kaiser, S. Manion, and A. Milliken of the JPL group for valuable discussions and help with the preparation of this article.

References

1   W. J. Kaiser and L. D. Bell. *Phys. Rev. Lett.* **60**, 1406, 1988.

2   L. D. Bell, W. J. Kaiser, M. H. Hecht, and L. C. Davis. *J. Vac. Sci. Technol. B.* **9**, 594, 1991.

3   L. D. Bell and W. J. Kaiser. *Phys. Rev. Lett.* **61**, 2368, 1988.

4   L. J. Schowalter and E. Y. Lee. *Phys. Rev. B.* **43**, 9308, 1991.

5   M. Prietsch and R. Ludeke. *Surf. Sci.* **251/252**, 413, 1991; *Phys. Rev. Lett.* **66**, 2511, 1991; R. Ludeke and M. Prietsch. *J. Vac. Sci. Technol. A.* **9**, 885, 1991.

6   J. L. Freeouf, T. N. Jackson, S. E. Laux, and J. M. Woodall. *J. Vac. Sci. Technol.* **21**, 570, 1982; R. T. Tung. *Appl. Phys. Lett.* **58**, 2821, 1991.

7   M. H. Hecht, L. D. Bell, W. J. Kaiser, and L. C. Davis. *Phys. Rev. B.* **42**, 7663, 1990.

8   L. D. Bell, M. H. Hecht, W. J. Kaiser, and L. C. Davis. *Phys. Rev. Lett.* **64**, 2679, 1990.

9   P. Niedermann, L. Quattropani, K. Solt, A. D. Kent, and Ø. Fischer. *J. Vac. Sci. Technol. B.* **10**, 580, 1992.

10 L. Quattropani, K. Solt, P. Niedermann, I. Maggio-Aprile, Ø. Fischer, and T. Pavelka. To be published in *Appl. Surf. Sci.*

11 S. J. Manion, L. D. Bell, W. J. Kaiser, M. H. Hecht, R. W. Fathauer, A. M. Milliken, and V. Narayanamurti. To be published.

12 H. D. Hallen, A. Fernandez, T. Huang, R. A. Buhrman, and J. Silcox. *J. Vac. Sci. Technol. B.* **9**, 585, 1991.

GEORGE N. MARACAS

The method of measuring free carrier (or doping) concentration profiles in semiconductors by capacitance–voltage (C–V) relies on modulating the depletion region in an MOS capacitor or field effect transistor (MOSFET), metal Schottky barrier, electrolyte–semiconductor contact, or p–n junction by an applied voltage. Superposition of a small ac voltage onto the dc bias allows the measurement of a capacitance of a thin region of doped material underneath a well-defined geometry contact (i.e., in which the area is precisely known). Scanning the reverse dc bias from zero to the breakdown voltage of the doped material provides a measure of the free carrier concentration versus depth. In simple structures such as Schottky barriers and p–n junctions the extraction of the doping profile is relatively simple. If all the dopant atoms are electrically active, then this information is obtained by analyzing the C–V characteristics, where the depletion depth is proportional to the area divided by the capacitance. The doping profile is proportional to $1/[A^2 d(1/C^2)/dV]$.

In the case of an MOS capacitor[1] the measurement is more complicated, in that special care must be taken since the device can be biased in accumulation, depletion, or inversion by different combinations of dc voltage and ac signal frequency. In the MOS capacitor case, extraction of interface state density and insulator mobile charge is also possible. The latter is achieved by controlling the mobile charge location with an applied dc bias and an elevated temperature stress cycle, which increases the mobile charge diffusion rate.

Commercial C–V measurement systems have a high degree of automation for material parameter extraction, including automatic bias scans and high temperature cycles. Contacting schemes have metal probes to contact pads, chemical contacts, and liquid metal contacts. The liquid metal contacting method is the mercury probe,[2] where liquid mercury produces a rectifying contact to the surface, obviating contact metallization patterning. Electrolytic C–V profilers[3] are available in which the rectifying barrier is produced by an electrolytic solution and the capacitance is measured at a constant dc voltage. The semiconductor is automatically and controllably etched to obtain depth information; thus, the limitation of depth from the breakdown voltage for a particular doping is relaxed.

Information	Free carrier concentration
Range of elements	Not element specific
Destructive	Yes, Hg probe no
Depth profiling	Yes
Depth resolution	Limited by Debye length

Depth limit	C–V breakdown voltage depth; electrolytic C–V no limitation
Doping range	$\sim 10^{14}$–10^{19} cm^3
Sample requirements	Depends on contacting scheme
Cost	$5000–$50,000 (depends on contacting scheme and automation level)

References

1   D. K. Schroder. *Semiconductor Material and Device Characterization*. Wiley-Interscience, New York, 1990.

2   P. S. Schaffer and T. R. Lally. *Solid State Tech.* **26**, 229–233, April 1983.

3   P. Blood. *Semi. Sci. Tech.* **1**, 7–27, 1986.

Deep Level Transient Spectroscopy (DLTS)   4

N. M. JOHNSON

Deep level transient spectroscopy (DLTS) is an experimental technique for characterizing the electronic properties of deep level defects in semiconductors. It is the most sensitive spectroscopy yet devised for the detection and characterization of such defects. The measurement is performed on semiconductor devices that possess a voltage-modulable space charge layer. The most commonly used devices are Schottky-barrier diodes, p–n junction diodes, and metal–insulator–semiconductor (MIS) or metal-oxide semiconductor (MOS) capacitors. For a given deep level defect, the DLTS measurement can, in principle, yield several phenomenological parameters that characterize the defect and its interaction with free charge carriers. These parameters include activation energies for emission of charge carriers, cross sections for the capture of free carriers, and the spatial (depth) distribution of the defect, the last parameter being a property of the material rather than of the defect. In combination with perturbation techniques (e.g., uniaxial stress) or complementary measurements (e.g., electron spin resonance), DLTS can be used for the microscopic identification of electronic defects.

Main Use	Electronic characterization of deep level defects in semiconductors
Sample requirements	Electronic devices with voltage-modulable space charge layers (e.g., Schottky diode)
Quantification	Activation energies for charge emission, cross sections for carrier capture, and defect concentrations
Sensitivity	Typically 10^{-4} of the doping concentration (e.g., 10^{11} defects/cm^3 for 10^{15} dopants/cm^3)
Depth probed	Space charge layer of device, upper limit set by reverse bias breakdown (e.g., ≤ 20 μm for 10^{15} dopants/cm^3 in Si)
Instrument cost	\$50,000–\$120,000
Size	Standard laboratory bench or equivalent floor space

Recommended Reading

Johnson, N. M. "Measurement of Semiconductor-Insulator Interface States by Constant-Capacitance Deep Level Transient Spectroscopy." *J. Vac. Sci. Tech.* **21** (2), 303–314, 1982. A review of the DLTS measurement and analysis for the determination of interface state distributions in MIS capacitors.

Johnson, N. M., D. J. Bartelink, R. B. Gold, and J. F. Gibbons. "Constant-Capacitance DLTS Measurement of Defect Density Profiles in Semiconductors." *J. Appl. Phys.* **50** (7), 4828–4833, 1979. Describes the constant-capacitance mode of the DLTS measurement and its formulation for determining depth profiles of deep levels.

Lang, D. V. "Fast Capacitance Transient Apparatus: Application to ZnO and O centers in GaP *p–n* Junctions." *J. Appl. Phys.* **45** (7), 3014–3022, 1974. The first of the two original articles introducing the DLTS technique.

Lang, D. V. "Deep Level Transient Spectroscopy: A New Method to Characterize Traps in Semiconductors." *J. Appl. Phys.* **45** (7), 3023–3032, 1974. The second of the two original articles introducing the DLTS technique.

Lefevre, H., and M. Schulz. "Double Correlation Technique (DDLTS) for the Analysis of Deep Level Profiles in Semiconductors." *Appl. Phys.* **12** (1), 45–53, 1977. Introduces the double-correlation DLTS technique for determining the depth distribution of bulk deep levels in semiconductors.

Miller, G. L., D. V. Lang, and L. C. Kimerling. "Capacitance Transient Spectroscopy." *Ann. Rev. Mat. Sci.* **7**, 377–448, 1977. An early review of the DLTS technique and its application to bulk defect characterization.

Pensl, G. "Measurement Methods." In *Landolt-Börnstein: Data Tables on Impurities and Defects*. Series III, Vol. 22, Part b (O. Madelung and M. Schulz, Eds.), Springer-Verlag, Berlin/Heidelberg, 1989, Chapt. 3. A recent comprehensive compilation of electrical measurements on semiconductors, including many variations of the DLTS technique, with an extensive list of references.

Williams, R. "Determination of Deep Centers in Conducting Gallium Arsenide." *J. Appl. Phys.* **37** (9), 3411–3416, 1966. Historically interesting as the first experimental demonstration of the detection and characterization of a deep level defect with a transient capacitance measurement.

Dynamic Secondary Ion Mass Spectrometry (D-SIMS)* 5

In Secondary Ion Mass Spectrometry, SIMS, the sample is bombarded by a primary ion beam (typically in the few 100 ev to many keV energy range). Several alternative ion species are used, depending on the requirements of the analysis, which causes atoms and clusters of atoms to be sputtered from the sample surface. These atoms and clusters consist largely of neutral species, but a small fraction of both positively and negatively charged species occur (the secondary ions). These secondary ions are passed into a mass spectromter (several different types are in use), where the number of ions at each mass/charge ratio are counted. The secondary ion counts may be turned into atomic concentartions present in the sample, if (and only if), comparison to standards of the same or very similar material is available. In Dynamic SIMS, the secondary ion signal intensities are monitored as a function of sputter time (depth), allowing concentrations to be determined as a function of depth.

The strength of Dynamic SIMS is the ability to to monitor for all atoms, including hydrogen, helium, and including isotopes, over a wide dynamic range of concentrations (down to ppb to ppt levels). A lateral resolution down to below 10 nm is possible, depending on the nature of the instrument and sample (not at trace ppb levels). A depth resolution of 2–30 nm is possible, depending on material, sputter conditions, and the depth sputtered. The first few nm depth sputtered, however, is rarely quantifiable, however, owing to the initial non-equilibrium nature of the sputtering process and the lack of appropriate standards for a region which can be substantially different from the bulk material.

The major use of Dynamic SIMS is the quantitative determination of the concentration of dopants in semiconductor material, as a function of depth, over a dynamic range of 5 orders of magnitude. It is also used in a similar way for trace impurities in metals and alloys and geological materials.

DAVID C. JOY

Charge collection scanning electron microscopy, often referred to as electron beam induced current (EBIC) microscopy, is a technique that permits the scanning electron microscope (SEM) to image individual junctions, devices, and electrically active defects in semiconductor materials. Quantitative information about parameters such as resistivity and minority carrier diffusion lengths can also be obtained.

Main Use	Determination of position and depth of junctions and electrically active defects in semiconductors
Sample requirements	All semiconductors to which ohmic contacts can be fabricated
Destructive	No, but possible damage to gate oxides from electron beam
Lateral resolution	Typically 0.5 µm—depends on beam energy
Depth probed	Up to about 10 µm in silicon at 30 keV beam energy
Depth profiling	Yes, by changing beam energy
Detection limit	Can detect doping effects at $<10^{16}/cm^3$, defects at $<10^4/cm^2$
Quantitative	Yes, can measure diffusion lengths and carrier lifetimes
Imaging	Yes, this is the standard mode of usage
Imaging/mapping	Yes
Instrument cost	See SEM
Size	See SEM

Physical Principles

When an electron beam impinges on a semiconductor, electron–hole pairs are generated, a process which requires an amount of energy e_{eh}, where typically e_{eh} is about three times the energy of the band gap—for example, e_{eh} for silicon is 3.6 eV. If we assume that all the energy deposited by the incident beam of energy E_0 is ultimately available for the generation of electron–hole pairs, then the number of carrier pairs formed, n_{eh}, will be

$$n_{eh} = E_0/e_{eh} \tag{1}$$

For example, a 10 keV beam incident on silicon could produce $10\,000/3.6 \approx 2800$ electron–hole pairs.

In the absence of any external excitation the resistivity of a semiconductor material is high because there are few, if any, mobile charge carriers; but when a beam is turned on, each incident electron produces electron–hole pairs, and these free charge carriers produce so-called b or beam induced conductivity in the material.[1] Because the electron and hole have opposite charges they are electrostatically attracted and will tend to drift together through the lattice, maintaining local electrical neutrality. After a short time the electron will fall back across the band gap and recombine with the hole, giving up the energy used to form the original carrier pair in the form of cathodoluminescent radiation. If, however, a potential difference is maintained across the semiconductor, then the electric field it generates will separate the electrons and holes, because the electrons will tend to move towards the positive end of the sample, while the holes will drift towards the negative end, and a current will flow which can be monitored in an external circuit (Figure 1*a*).

In practice, a more useful procedure is to generate the field internally within the semiconductor. For example, the depletion region extending a few micrometers on either side of the physical location of a *p–n* junction contains a field of several thousand volts per centimeter (Figure 1*b*). If the incident electron beam is placed on the specimen in either the *p*-type or *n*-type regions, well away from the depleted zone, then although electron–hole pairs will be generated, the material in which they are produced is electrically neutral and has no field across it, so they will

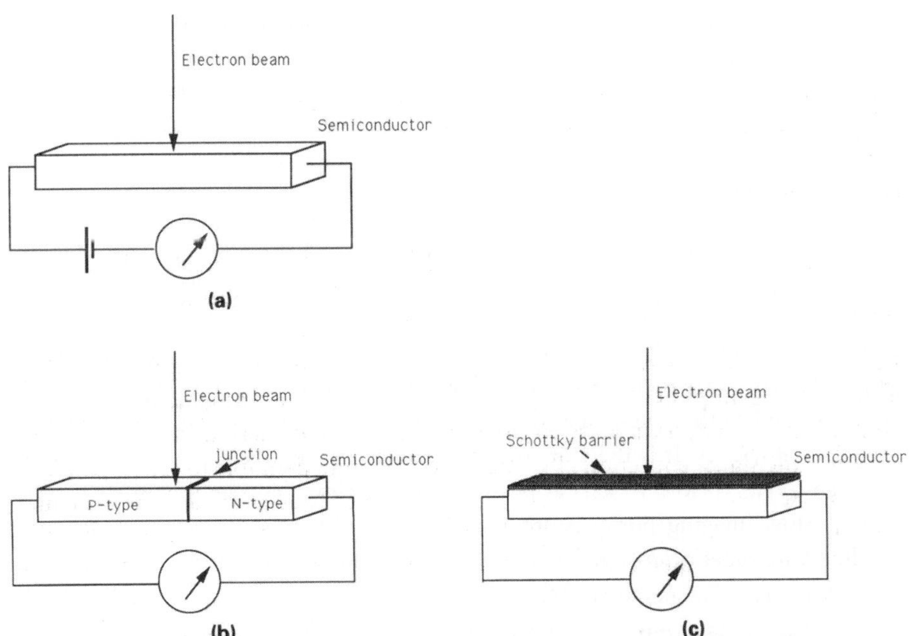

Figure 1 (*a*) ß or beam induced conductivity in a semiconductor irradiated by an electron beam. (*b*) Charge collection or electron beam induced current (EBIC) imaging with a *p–n* junction. (*c*) EBIC imaging with a Schottky barrier.

recombine and no external effect will be observed. However, if the beam is placed in the junction region, the carriers produced will see the depletion field, the carrier pairs will be separated, and the motion of these charges within the specimen will produce a flow of current I_{cc}—the charge collected or EBIC—in the external circuit given by the relation

$$I_{cc} = n_{eh}\, I_B = I_B\, E_0 / e_{eh} \tag{2}$$

since each incident electron can produce n_{eh} carrier pairs. I_{cc} is seen to be substantially greater (100× to 1000× or more) than the incident current I_B. If the beam is put close to, but not actually inside, the depleted zone, some fraction of the carriers will still diffuse back into the depletion region and be collected by the field, producing a current that will vary as $\exp(-\Delta s/\lambda)$, where Δs is the distance of the beam from the edge of the depletion zone and λ is the minority carrier diffusion length. If the incident beam of electrons is scanned, as in the SEM, then the current flowing around the loop can be amplified and displayed as a function of the beam position, producing a charge-collected, or EBIC, image. In the case of the *p–n* junction, the signal would show a maximum over the depletion region surrounding the junction and then fall back to zero on either side over a distance determined by the minority carrier diffusion length and the incident electron range—typically a region a few micrometers wide.

An alternative geometry (Figure 1*c*) is to place a Schottky barrier on the surface of a semiconductor. The depleted region extending downwards from the Schottky is again associated with a field which can separate the electron–hole pairs and given current flow in the external circuit. In this case a strong signal will be collected over the entire lateral extent of the Schottky barrier allowing, for example, for the low magnification examination of a wafer.

Practical Details

A typical imaging current, I_B, for an SEM is about 10^{-9} A at 20 keV; so from Equation 2 it can be predicted that the charge collected circuit, in silicon, would be ~0.5 µA. Measuring a current this large is certainly not a problem. However, the predicted current is actually a short-circuit current, and if the input impedance of the imaging amplifier used is R_L, then the ohmic voltage drop across the input, $I_B \cdot R_L$, is in the opposite sense to the potential at the depletion layer and, hence, will effect the signal collection efficiency. It is therefore desirable to use an amplifier with as low an input impedance (e.g., below 10 kΩ) as possible. The specimen current amplifiers used for normal SEM imaging purposes are adequate, but current-sensitive instrumentation amplifiers are superior. A signal bandwidth of 30–100 kHz is required in the amplifier for satisfactory visual rate imaging.

Electrical contacts must also be made to the material to be imaged. These contacts must be ohmic and low in resistance. This is no problem for silicon, but for some III–V and II–VI materials fabricating suitable contacts may be very difficult. As a general rule, satisfactory EBIC imaging will not be achieved unless the material

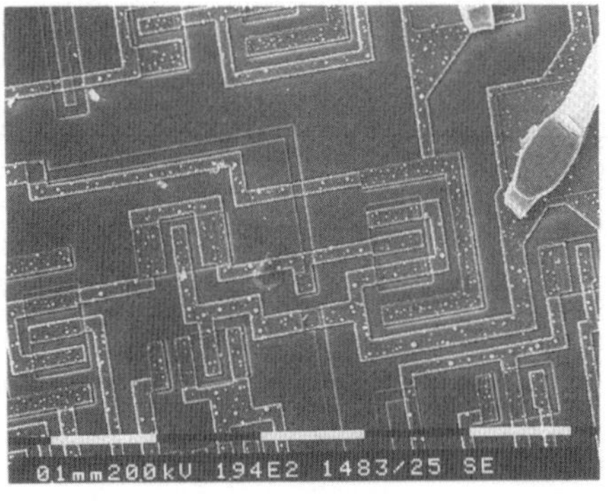

(a)   (b)

Figure 2   Secondary electron topographic image (*a*) and EBIC image (*b*) of the same area. Contrast in the EBIC image comes from current collected at junctions within the device.

or device under test shows a reverse-to-forward resistance ratio of at least ten to one when measured by a conventional volt-ohmmeter placed across the signal leads.

Applications

The original use of EBIC imaging was to observe *p–n* junctions in devices.[2] Satisfactory images can normally be obtained by collecting the signal across the $+V_e$ and $-V_e$ power rails of the device, since this will give a dc path to every junction. Compare Figure 2*a*, the normal secondary electron topographic of a device, with Figure 2*b*, the EBIC image. The junction appears as black or white, depending on whether the *p* or the *n* side of a given junction is connected to the positive-going signal lead. Junctions can be imaged at depths beneath the surface up to the maximum range of the incident electron beam—in silicon, a depth of 0.5 µm at 5 keV beam energy and 5 µm at 20 keV beam energy. If the incident beam energy is gradually increased on a given area, the energy at which a junction is first visible can be used to estimate its depth. It is not normally necessary to remove surface passivation films in order to perform EBIC imaging, but the presence of this layer will reduce the penetration of the beam into the device proper.

The Schottky barrier geometry is best suited to the study of materials in the unprocessed, or partially processed, state. In principle, the barrier regions could be as large as required to cover the area of interest, but because the capacitance represented by the Schottky diode is in parallel with the amplifier input impedance, too large a diode will give a long time-constant and require very slow scanning for a satisfactory image. It is usually better to cover the area of interest with an array of small (250–500 µm diameter) barrier dots and select the region required by moving the top contact wire to the appropriate dot. Typical contrast effects include:

1 Variations in the depletion depth due to changed resistivity (due to planned or random changes in doping) will produce corresponding changes in the collected

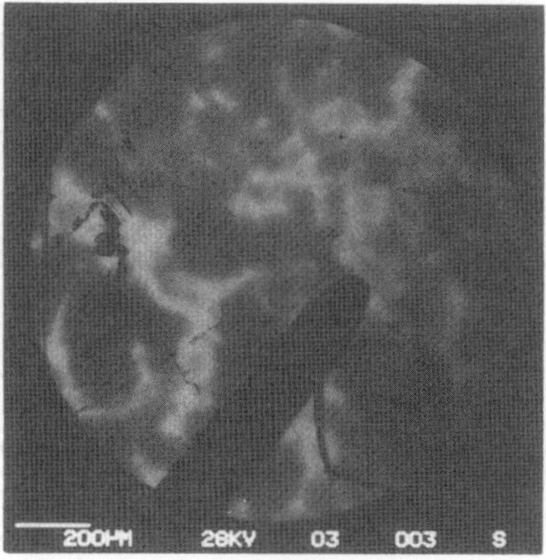

Figure 3   EBIC image of silicon wafer. The circular area shows the extent of the Schottky barrier, and the shadow is from the spring-loaded top-contact wire to the sample. The dark areas in the image correspond to regions of low depletion depth where impurities have lowered the resistivity.

signal. For example, the darker, cloud-like regions visible in Figure 3 show regions where the resistivity of a silicon wafer has been changed by impurity concentrations at the $10^{16}/cm^3$ level during crystal growth.

2 Changes in Schottky barrier height will also modulate the signal collection. For example, the A and B forms of cobalt silicide have barrier heights differing by 0.1 eV, so images of silicon formed with a mixed cobalt silicide barrier show a characteristic mottled appearance.

3 Individual electrically active defects (dislocations and stacking faults), lying within the beam range from the surface will allow the electron–hole pairs to recombine locally and produce a decrease in the collected signal.[3] Figure 4 shows such defects in a polycrystalline solar cell. The width of the defect image depends

Figure 4   EBIC image of dislocations in a polycrystalline solar cell. In this case the defects are nearly end-on to the surface.

on the energy of the beam and the depth of the defect and is typically ~1–3 µm; so it is easy to see individual dislocations even in a low-magnification (100×) image. The EBIC method is therefore well-suited to the examination of materials in which the dislocation density is very low ($<10^4$/cm^2).

4 Avalanche breakdown sites at impurities may produce intense and unstable signal effects under conditions of high incident current.[4]

Although it is the imaging aspects of charge collection microscopy that have been considered here, over 90% of the published literature in this field discuss the use of EBIC to determine minority carrier diffusion lengths and lifetimes. A wide variety of techniques have been described that permit spatially resolved measurements to be performed on a 10 µm scale. For a more complete discussion, see References 4 and 5.

References

1 K. G. McKay. *Phys. Rev.* **74**, 1606–1621, 1948.

2 T. E. Everhart, O. C. Wells, and R. K. Matta. *Abs. Electrochem. Soc.* **12**, 1–5, 1963.

3 J. J. Lander, H. Schreiber, T. M. Buck, and J. R. Mathews. *Appl. Phys. Lett.* **3**, 206–208, 1963.

4 H. J. Leamy. *J. Appl. Phys.* **53**, R51–80, 1982.

5 D. B. Holt and D. C. Joy. *SEM Microcharacterization of Semiconductors.* Techniques of Physics Series, Vol. 12, Academic Press, London, 1989.

Energy-Dispersive X-Ray Spectroscopy (EDS)* 7

Energy-Dispersive X-Ray Spectroscopy, EDS, is a specific technique for the detection and energy distribution determination of X-Ray Fluorescence, XRF. XRF is the phenomena where X rays are emitted from a material when bombarded by high energy radiation (electrons, ions, X rays, neutrons, gamma-Rays). Some of the X-ray energies emitted are characteristic of the atoms present, allowing atomic identification in the material of interest.

In EDS a solid state X-ray detector, usually lithium drifted Si, and pulse counting electronics are used. The detector converts an incoming X-ray photon into an electronic pulse of amplitude proportional to the energy of the X ray. The signal processing electronics counts the number of pulses of each different amplitude, resulting in a histogram of X-ray energy versus intensity. The X-ray energies allow atom identification, and the relative peak intensities can be related to relative atomic concentrations by comparision to standards, or by theoretical calculations.

All elements with an atomic number higher than Li are, in principle detectable, though effectively dealing with low Z elements requires some care and correctly set up instrumentation. In practise, EDS is primarily used in conjunction with e-beam columns (SEMs, TEMs, STEMs, Auger instruments) as the excitation source. The depth probed is dependant primarily on the energy of the electron beam, and on the material being probed. It can vary from as little as 20 nm (high Z material, low beam energy) to as much as 5000 nm (low Z material, high beam energy). The lateral resolution is similar to the depth probed but is not determined by the primary beam diameter. It is determined by the beam energy and the sample material, because the beam spreads by scattering as it penetrates into the material, creating X rays in this enlarged volume.

The particular strength of EDS is the simultaneous (or parallel) detection of elements rapidly and cheaply, by placing the physically small detector inside the SEM, or other electron beam system. As such it adds an elemental analytical capability to imaging electron beam columns. Its drawback is that the typical solid state detector has very poor energy resolution, which means that there are significant peak overlaps, sometimes making it difficult to distinguish elements. This is particularly true in the low energy region.

Focused Ion Beams (FIBs) **8**

JON ORLOFF

Introduction

Focused ion beams (FIBs), in the present context, means ion beams with a focused spot size of less than 1 μm with a current density in the focused spot $J \approx 1$ A·cm^{-2}. Only in the past 15 years have such capabilities in an FIB become possible, making the FIB tool extremely useful. FIB tools are now widely used for micromachining, failure analysis, and quality assurance in the integrated circuit fabrication process, as well as for surface analysis, lithography, direct implantation, etc. In fact, in the first two of these applications, FIBs are proving to be indispensable for semiconductor device manufacturing. FIBs have recently achieved this impressive performance as a result of the harnessing of field emission ion source technology. Prior to the use of field emission technology, conventional ion sources such as duo-plasmatron sources were used, and sub-micrometer beam sizes were possible[1] only with $J << 1$ A·cm^{-2}. Current FIB capabilities translate to a micromachining material removal rate of ~1 μm^3·nA^{-1}·s^{-1}. This makes it feasible for FIBs to be utilized for integrated circuit dissection on a micrometer scale, for failure analysis, and for integrated circuit modification and repair.

Principles and Instrumentation

An FIB system consists of a field emission liquid metal ion source (LMIS) coupled with an electrostatic optical column, which focuses and controls the ion beam. The technology developed from efforts in the early 1970s[2–5] to develop high performance systems using gas-phase field ionization sources (GFISs). In the mid-1970s, research on LMIS began in earnest as it was realized that such sources have many very attractive properties, including the ability to operate at or near room temperature, high intensity, and very low noise.[6] LMIS were first used in an FIB system in 1978[7] and were rapidly employed for a wide variety of applications thereafter.

The heart of an FIB system is the LMIS, which generally consists of a needle with an end radius of ~10 μm coated with the liquid metal from which ions are to be produced. The needle is located in close proximity to an electrode called an extraction electrode, which creates a high electric field when a high voltage is applied to the needle relative to the electrode. The balance of forces on the liquid due to the electrostatic field stress and the surface tension force of the liquid cause the liquid to assume a conical equilibrium shape, which is called a Taylor cone.[8] Both the electrostatic force and the surface tension force are proportional to the square of the inverse of the cone radius of curvature. As the radius of the end of the liquid approaches ~5 nm, the electric field becomes high enough to cause field evaporation and ionization of the evaporated atoms. At this point a flow of liquid metal through the cone-shaped structure is established which replaces the atoms lost through ionization. Stable ion currents can be produced in the range ~1–50 μA.

LMIS are characterized by low noise, angular intensities $I' \approx 20$ mA·sr^{-1}, long life, and very stable operation. The properties required by the liquid metal to be ionized are low vapor pressure at the melting point, high surface tension, and modest chemical reactivity with the substrate needle supporting it.[9] Many metals and some non-metals have been used for LMIS operation, including Al, As, Au, B, Be, Bi, Cs, Cu, Ga, Ge, Fe, In, Li, P, Pb, Pd, Si, Sn, U, and Zn. Ga is the most popular element for micromachining, scanning ion microscopy, etc., because of its low melting point, extremely low vapor pressure, and general ease of handling. Elements that have a high vapor pressure at the melting point can be used in the form of eutectic alloys if a mass filter is available in the optical system.

Because the area at the end of the liquid cone where ionization takes place is very small, the current density is extremely large and, consequently, space charge effects are significant. The primary effects are an increase in the virtual source δ size from $\delta \approx 1$ nm for a GFIS to $\delta \approx 50$ nm for an LMIS and an increase in the fwhm of the energy distribution ΔE from $\Delta E \approx 1$ eV fwhm for a GFIS to $\Delta E \geq 5$ eV for an LMIS. In addition, ΔE increases very rapidly with the total current I drawn from an LMIS, when I is greater than ~2 μA,[6] while the angular intensity increases relatively slowly.

FIB instrumentation has developed along two lines. Low energy systems (E_{beam} < 35 keV) have been developed in a number of university and industrial research laboratories, whereas high energy systems (E_{beam} > 100 keV) equipped with mass filters and designed for direct implantation and lithography were developed in a few industrial laboratories. The latter instrumentation is quite expensive, and no commercial development of them is taking place at the present time (1993); rather, the emphasis is on low energy systems with pure metal sources to be used primarily for micromachining, microscopy, and surface analysis (high spatial resolution SIMS). These low energy systems have demonstrated imaging resolution of <50 nm and are routinely used for micromachining at < 100-nm beam size. Under pressure from the semiconductor device industry, many additional capabilities have been added to commercial FIBs, including SIMS, chemically enhanced etching capability, high precision specimen stages, metal deposition, light microscopy, and high resolution SEM capability.

Images are produced in an FIB instrument with secondary electrons in the same way as in a scanning electron microscope. The yield of secondary electrons is high, of the order of one electron per incident ion, and it is strongly dependent on topography and crystalline structure. The ion beam also sputters atoms from the surface being analyzed and some of those sputtered atoms are ionized. Therefore, it is possible to image the specimen with secondary ions as well as with secondary electrons.

Applications of FIBs

The main applications of FIBs are in failure analysis and circuit modification and repair in the semiconductor manufacturing industry. These applications require modest energy beams (20–30 keV), so the FIB optical columns are relatively simple. By using the micromachining capability of a submicrometer FIB it is possible to dissect a small region of a circuit to examine, for example, quality of metallization and to search for contaminants without the need to subject the device to major destructive surgery, such as chemical stripping of whole layers of material.[10] Suspicious areas in a circuit, which can be identified by analysis of device performance or by peculiar morphology, are located by means of an SEM. These areas can then be analyzed by sputtering away only cubic micrometers of material and using the FIB or an SEM to inspect the subsurface regions. Circuits can be modified or rewired by cutting conducting runs and adding new conductors. The latter can be done by decomposing an organometallic gas with the ion beam, causing it to deposit a metallic conducting layer.[11] In order to perform these applications, it is necessary that the FIB system have an accurate and precise stage ($1:10^5$ precision) and that it be controlled by a digital computer, which directs the beam position and blanks and unblanks the beam appropriately and drives the stage.

Secondary ions are produced by the focused ion beam when material is sputtered from a surface; these secondary ions can be collected and analyzed, which is the basis for SIMS. Only a small fraction of the sputtered atoms are ionized in the process. SIMS has the advantages of great chemical selectivity and sensitivity to only the surface of a specimen. Since ions are normally produced when the focused ion beam strikes a specimen, it is natural to include SIMS as one of the capabilities of a FIB system. It is particularly useful for end-point detection and can be used, for example, when milling through a layer of passivation while seeking a conducting run in a circuit. It is also useful for producing chemical maps of specimens and for identifying small contaminating particles buried inside an electronic circuit. Since the FIB is rastered, gating techniques can be used and dynamic SIMS analysis is possible (see Technique Summary 5, Dynamic SIMS). SIMS is useful for end-point detection when milling through layered structures and for making elemental maps and analyzing small (< 1 μm) particles in semiconductor devices.

The rate of material removal with an FIB can be increased several-fold by using chemical assistance such as Cl_2 or I. When an appropriate etching chemical is sprayed on a specimen surface while machining with the FIB, it causes surface material to be removed mainly in a volatile form.[12, 13] The beam induced chemical reactions result in a much higher removal rate than can be achieved with the ion beam by itself. In addition, higher quality machined surfaces can be obtained because redeposition of sputtered material on freshly machined surfaces is minimized.

When an FIB optical column is added to a high resolution SEM, the resulting dual-beam system can be a very powerful tool for inspection. Such a system, if equipped with a field emission electron beam column, can be used for nondestructive imaging at very high resolution (<5 nm), in addition to exploiting all of the other capabilities of the FIB system. With a proper specimen handling stage it is possible to use either beam with little compromise in performance and to simultaneously perform micromachining while observing the results with the SEM. First invented in 1984,[14] dual column systems became commercially available in 1992.

Present FIB equipment costs from ~$30,000 for a stand-alone focusing column to ~$1,000,000 for the most complete system, including the various features described above. For further information on LMISs and FIBs, the reader is directed to References 15 and 16.

References

1 A. R. Hill. *Nature*, **218**, 292, 1968.

2 R. Levi-Setti. *Scanning Electron Microscopy*. **125**, 1974.

3 W. Escovitz, T. Fox, and R. Levi-Setti. *Proceedings*. Annual Meeting of the Electron Microscopy Society of America. **33**, 304, 1975.

4 J. H. Orloff and L. W. Swanson. *J. Vac. Sci. Tech.* **12**, 1209, 1975.

5 J. Orloff and L. W. Swanson. *Scanning Electron Microscopy*. **10**, 57, 1977.

6 L. W. Swanson. *Nucl. Inst. Meth.* **218**, 347, 1983.

7 R. L. Seliger, J. W. Ward, V. Wang, and R. L. Kubena. *Appl. Phys. Lett.* **34**, 310, 1979.

8 G. I. Taylor. *Proc. Royal Soc. A.* **280**, 383, 1964.

9 M. J. Bozack, L. W. Swanson, and J. Orloff. *SEM/1985/IV.* 1339, 1985.

10 R. Boylan, M. Ward, and D. Tuggle. *Proc. Int'l. Sym. for Testing and Failure Analysis*. ASM International, Materials Park, OH, 1989, p. 249.

11 A. D. Dubner, G. M. Shedd, H. Lezec, and J. Melngailis. *J. Vac. Sci. Tech.* **B5**, 1434, 1987.

12 Y. Ochiai, K. Gamo, and S. Namba. *J. Vac. Sci. Tech.* **B3**, 67, 1985.

13 A. Gandhi and J. Orloff. *J. Vac. Sci. Tech.* **B8**, 1814, 1990.

14 P. Sudraud, G. BenAssayag, and M. Bon. *Microelectron. Eng.* **6**, 583, 1987.

15 R. A. D. Mackenzie and G. D. W. Smith. *Nanotechnology*. **1**, 163, 1990.

16 J. Orloff. *Rev. Sci. Instrum.* **64**, 1105, 1993.

In Infrared Spectroscopy (IR), a beam of light (electromagnetic radiation) in the infrared wavelength region impinges on the sample and the wavelength/frequency is scanned. Whenever there is a match of the light frequency to the vibrational frequency of a vibrational mode within the sample (i.e., vibration of atomic positions) it is possible that the radiation at that frequency will be absorbed and the vibration excited (selection rules apply and not all vibrations can be excited). The absorption occurring, as a function of IR wavelength, is monitored by comparing the input intensity of the radiation to the output intensity, thus revealing the vibrational frequencies existing in the sample. Since a vibrational frequency value is dependant on the bonding between the atoms involved in that vibration, vibrational frequencies of materials are characteristic of the chemical groups existing in the material (e.g., OH, CN, CH_2, COOH, etc.), so determination of the vibrational frequencies present allows determination of which chemical groups are present. With standards the concentrations of the groups present can also be determined from the strength of the absorptions.

Fourier Transform IR simply describes the most common modern technique for scanning the wavelength of the impinging IR beam. It is based on the Michelson interferometer, where constructive and destructive interferences between the two halves of a split light beam are controlled by changing the path length of one of the beams with respect to the other by moving reflecting mirrors. Since the path length difference for destructive/constructive interference depends on wavelength, the oscillation of the mirrors provides a way of scanning the wavelengths from a polychromatic light source.

For solids, IR spectroscopy can be performed either in reflection or in transmission if the sample is thin enough to pass the radiation. This varies enormously with material because absorption coefficients vary enormously (some materials are transparent to IR, some highly absorbing). In reflection, coupling the radiation to an optical microscope allows a lateral resolution down to 20 µm, with the depth probed depending on absorption coefficient, but usually being many µm. Another reflection mode is Attenuated Total Reflection, ATR, where light enters at grazing angle and probes only the surface region (10's of nm). IR of solids does not require a vacuum, but the path length of the radiation through ambient atmosphere must be purged using dry nitrogen, to avoid gas phase absorption from strong absorbers such as water vapor.

The strength of IR is the qualitative and sometimes quantitative (with standards) identification of the presence of chemical species, or functional groups, sometimes down to trace levels in liquids and solids. For solids stress, strain, crystallinity/ amorphousity, and inhomogeneities can be detected from absorption peak broadenings and shifts.

Hall Effect Resistivity Measurements 10

GEORGE N. MARACAS

Resistivity and carrier concentration and mobility are easily obtained by Hall effect measurements. Temperature-dependent Hall measurements provide carrier compensation ratio, total impurity density, ionization energy, and dominant scattering mechanism. Extending the technique to high variable magnetic fields (~12 T) allows magnetoresistance characterization of mobile carriers in two-dimensional conducting channels and of quantized state energies.

Samples require a well-defined conducting-layer thickness or conducting channel. Isolation of the channel from the substrate can be achieved by ion implantation or epitaxial growth. No isolation is required if bulk material (e.g., substrates or thick slabs) is to be characterized. Ohmic contacts are placed onto patterns such as a square slab, Van der Pauw (VDP or clover-leaf) pattern, or Hall bar. The square slab is the easiest, the VDP has the best contact placement error immunity, and the Hall bar is the most appropriate for magnetoresistance measurements. Sample dimensions are on the order of 1 cm.

The Hall effect technique uses a four point probe geometry and exploits the Lorentz force behavior of mobile carriers in an applied magnetic field to extract the "Hall coefficient," R_H, which is directly related to the carrier density. By combining concurrent resistivity measurements on the same sample, extraction of free carrier concentration, type, and mobility are possible. In the simplest method, a constant current is applied between two contacts and the voltage is measured across the other two. Application of a magnetic field (~2 kG) creates a charge density imbalance that produces a voltage across the two contacts located perpendicular to the current flow direction. The magnitude of the voltage is directly proportional to the mobility; the polarity indicates the carrier type.

Mobility depth profiling can be accomplished by either including a rectifying "gate" or by chemical etching, which controls the conducting layer thickness between the contacts. Numerical techniques are required to extract the uniform or non-uniform doping/mobility profiles versus depth.

Commercial Hall effect systems automatically control the multiple contact terminal switching sequence and magnetic field application. The cost of an electromagnet raises the price of systems; some are available with small fixed magnets. Electrometers or very high impedance buffers are required to measure semi-insulating materials, which also raises the price of a system. Temperature-controlled cryostats (4 K to above room temperature) are also available and vary considerably in design and cost.

Information	Carrier concentration, mobility, type
Element range	Not element specific

Destructive   Yes

Depth profiling   Yes, limited

Depth resolution   Limited by Debye length

Depth limit   Barrier breakdown voltage depth; uniform etch depth
                               limit

Doping range $\sim 10^{14}$ to high 10^{19} cm^{-3}

Sample requirements   Four contacts, ~ 1 cm sample size

Cost   \$10,000–\$50,000 (depends on magnet size and type and
                               buffer impedance); high field magnets: $\sim$\$35,000

Recommended Reading

Blood, P., and J. W. Orton. *The Electrical Characterization of Semiconductors: Majority Carriers and Electron States.* Academic Press, New York, 1992, Chapt. 4.

In Inductively Coupled Plasma Mass Spectrometry (ICPMS), the sample is introduced by one of several ways into an inductively coupled plasma which fragments and ionizes all species present down to the atomic level. The ions are passed into a mass spectrometer (several types are in use) and mass analyzed to reveal which atoms are present. The plasma is highly efficient, and, coupled with an efficient and high resolution mass spectrometer, detection limits down to sub ppb are achievable for many elements. If a low resolution mass spectrometer is used, mass interferences limit the detection limit for some elements, notably Fe. Rough quantification to the 5 to 20% level is easy using standard sensitivity factors determined for the mass spectrometer. For better quantification standard solutions using a material matrix as similar to the sample as possible are used. Sub 1% accuracy can be achieved by using the method of isotope spiking.

The normal method of sample introduction is to nebulize an aqueous solution into the plasma. For solids, dissolution by a solvent, or digestion by an acid (normally nitric acid), is the preferred method, but direct sampling using laser ablation is possible. This method is the only way to achieve any spatial resolution. Laser ablated craters are typically 10's of μm wide and a fraction to a few μm deep.

The main uses of ICPMS are in two areas. One is the routine determination of contaminant levels in solutions and solvent, particularly water supplies. The second is for determination of trace levels of elements in semiconductors and alloys.

The visible light microscope dates back over 300 years for the study of natural materials (plants, animals). At its simplest level the method involves visible observation using white light, with useful magnification of up to 1400X. This allows observation resolving features down to around 0.2 µm, as opposed to about 200 µm by the naked eye. Resolving power can be increased by using UV radiation (doubled) or by immersing optics in oil, but both procedures greatly increase the compexity of instrumentation. For biomedical samples and many other materials (e.g., fibres, wood, hairs, pollens), the morphology revealed in the image is the main information, and materials identification can usually be made through use of extensive atlases of every conceivable structure observed. For materials science, simple morphology alone is often insufficient for identification, and it is necessary to investigate a number of other parameters such as opacity, color, refractive index, crystal system, fluorescence, and many others. These can be quickly determined using a variety of procedures and attachments to a microscope or several microscopes.

The main role of the light microscope is to image structures, either natural or man made, which are not observable to the naked eye, but have features greater than 0.1 µm. It is extensively used during the preparation of samples for observation by other analytical techniques, such as SEM or AFM, and should always be the first step in a materials analysis.

Low-Energy Electron Diffraction (LEED) is a technique for providing a two-dimensional diffraction pattern from a surface. It is only applicable to a surface with long-range order, typically a single crystal surface, or an adsorbed over layer. The method involves impacting a collimated beam (typically 0.1 mm diameter) of monoenergetic electrons, in the energy range 10 to 1000 ev (equivalent to a photon wavelength of 0.4 to 4 Å) onto the surface at normal incidence. Electrons in this energy range have very short inelastic mean free path lengths, so the electrons which are elastically diffracted back from the sample only come from the first few Å depth. They are separated from inelastically scattered electrons, which have lost energy, by a retarding grid system and are then detected on a phosphor screen. The observed diffraction pattern reveals the 2D unit cell type and dimensions. Determination of the actual surface atomic atom positions relative to each other, or the underlying layers, requires measurement of the diffraction intensities as a function of varying the electron energy and comparison to theoretical calculations on trial structures. The shapes, splittings, and broadening of the diffracted beams also carries information on surface disorder.

The major use of LEED has been to determine surface structures of both clean bulk crystalline material and adsorbed layers on such surfaces, by comparison to theory. It is also used routinely and simply to monitor the surface cleanliness of single crystal surfaces, and how well ordered the surface is (sensitive over about a 200 Å range in most instruments). A very specialized version exists, the Low Energy Electron Microscope, LEEM, which functions just like an imaging TEM and has a surface imaging resolution of 150 Å.

Neutron Activation Analysis (NAA)*   **14**

Neutron Activation Analysis (NAA) is a technique for trace analysis in bulk material. It involves two separate steps. The sample is initially irradiated by neutrons in a nuclear reactor to create the radio isotopes (which may take 1 to 7 days). This is followed by removal to a gamma-ray spectroscopy facility, where the gamma-rays given off during subsequent decay of the radio isotopes can be monitored. The gamma-ray energies emitted are characteristic of the atoms decaying and are determined using a solid state energy dispersive detector in an analogous manner to the X-ray analysis in EDS (see appendix 7), except that the energies are in the MeV range. The technique is a bulk one, since both neutrons and gamma-rays penetrate deeply. There is also no spatial resolution.

Both the initial neutron capture probability (cross-section) to create the radio isotopes, and the half-lives for the subsequent decay process vary enormously, in no particular pattern, across the periodic table. In fact, only about two thirds of the elements can produce radio isotopes this way. So trace element sensitivity varies enormously among elements, and also with the matrix, which may give off competing gamma-rays. For trace metals in Si or SiO_2, ppb to ppt sensitivity is easily obtained, but other semiconductor matrices such as GaAs are less favorable. Trace elements in plastics and biological materials can be measured because the matrix elements of C, O, H and N all have low neutron capture cross-sections. The material may suffer radiation and heating damage in the process, however.

The major uses of NAA are for trace determination of bulk impurities in semiconductor, biological, geological, and envonmental samples, and in forensic science. Quantification can be achieved from signal intensities using standards, or from calculations if enough information about the instrumetal conditions is available. The major drawbacks are the need to use reactor and certified gamma-ray facilities, the long time involved for an analysis, and the special handling needed for the samples.

Optical Scatterometry* 15

Optical Scatterometry determines the angular-resolved distribution of light (usually a visible wavelength laser) scattered from a surface. The distribution is a function of the properties of the bulk material (refractive index, homogeneity, orientation, density), the surface roughness, the surface contamination, and the source polarization and angle of incidence. The ratio of the scattered intensity to incident intensity as a function of scatter angle is the parameter determined.

On nominally smooth surfaces, measurement at a specific angle corresponds to determining the root mean average roughness at a particular spatial frequency (the intensity of scattered light is proportional to the roughness amplitude). Thus both long range and short range roughness can be determined, down to the A level, and with a spatial resolution of half the wavelength of the laser light.

Periodic structures (e.g., lithiographic etched lines) may also be characterized. Light is scattered into different diffraction orders, with the intensities in the different orders being strongly affected by the shape of the periodic struture (e.g., depth of etch, slope of sidewall).

The major use of Scatterometry is to characterize the surface roughness of nominally smooth surfaces and as a process metrology tool for lithography in wafer processing. It is fast, non-destructive, and can yield quantitative results when coupled to theoretical modelling. The disadvantage is that it gives an average picture and cannot observe any individual feature, unlike techniques such as SEM or AFM.

In Photoluminescence (PL), light in the visible wavelength region (usually laser light) is used to induce excited electronic states in the material concerned, and the light intensity then re-emitted as these states relax is monitored, usually as a function of wavelength, by passing through a lens and an optical spectrometer onto a photodetector. Another option is just to collect the total light emmitted, and a third option is to vary the wavelength of the light doing the excitataion.

The energy (wavelength) of a photoluminescence peak in the spectrum of intensity versus wavelength is determined by the difference between the enery of the excited state and the ground state to which it relaxes. Unlike X-ray fluorescence, where the energy levels involved are atomic in nature, in photoluminescence the energy levels involved are, for solids, the conduction bands (excited states) and valence bands (ground states), which are characteristic of the bonding effects in the material but not the individual atoms. No direct atomic identification is given, but peak energies relate to such properties as band gap, chemical compound, defect and impurity electronic states present, disorder and strain. It is usually qualitative rather than quantitative, though by comparison to standards it is possible to quantify trace impurity states. The depth probed varies from about 0.1 µm to several µm, depending on the light absorbtion properties of the material. Using a microscope, spatial resolutions down to the 1–2 µm range are possibe.

Major practical uses of PL are for determining band gaps in films and general defect detection. Impurity levels down to ppt levels can sometimes be detected. The mapping of the PL of wafers can be performed as a quality control method, particularly with respect to defects.

In Raman spectroscopy, a light beam of fixed wavelength (usually in the visible range) from a laser undergoes inelastic scattering as it interacts with the sample material. The photons can suffer energy loss if they excite vibrations in the sample. The energy losses, and therefore the vibrational frequencies involved, are determined by measuring the wavelengths of the scattered light. The value of an excited frequency is sensitive to the bonding between the atoms involved in that particular vibration.

Raman spectroscopy is closely related to infrared spectroscopy, which also determines the values of vibrational frequencies (see the FTIR summary), but it differs in the physics in that it involves a photon scattering process with energy loss, where as IR involves varying the photon energy and observing absorption occurring at the vibrational frequencies. Raman spectroscopy is better suited to optical microscopes, because of the single laser wavelength involved, and spatial resolution down to the one µm range is possible (cf. approximately a 10 µm limit for IR).

The depth probed depends very strongly on the optical properties of the material involved and the laser wavelength used. Moving into the UV range greatly reduces the probing depths, and allows a depth profiling capabilities if several wavelengths are used.

The major area of application for solids and liquids is chemical fingerprinting and the identification of unknown compounds. For solids, Raman is also used for phase identification, following amorphous/crystalline transitions, measurement of stress and strain, and, in the microscope mode, the detection and analysis of defects, including particles during wafer processing.

Reflection High-Energy Electron Diffraction (RHEED)*   18

In Reflection High Energy Electron Diffraction (RHEED), a beam of electrons in the 5 to 50 KeV energy range strikes the surface of a flat solid sample at grazing angle (1–5 degrees). The wavelength of these electrons is shorter than interatomic distances, and so diffraction of the beam takes place if there are ordered rows of atoms present at the surface of the solid. The diffracted beams are scattered into directions which depend on the spacing between rows (cf. LEED or X-ray diffraction), and the subsequent diffraction pattern is observed on a phosphor screen. The symmetry in the RHEED pattern reveals the symmetry of the atomic rows causing the diffraction, and the spacings in the pattern reveal the lattice spacings between rows. Broadenings and splittings of the diffraction features can also provide information on the degree of lattice disorder at the surface.

Owing to the grazing incidence, only the first few atomic layers of the surface are probed, despite the relatively high electron energies involved. Owing to the requirement for grazing incidents, very flat surfaces are needed (e.g., Si wafers), and there is very limited lateral resolution (typically 0.5mm by 4mm beam area on the surface).

The main use of RHEED is to monitor surface crystal structure during thin film epitaxial growth on single crystal surfaces. It is very sensitive to defects, and can distinguish 2D defects from 3D island growth.

In Rutherford Backscattering Spectrometry (RBS), a solid is bombarded with a monoenergetic beam of ions (typically He or H) in the one to several MeV energy range. As the beam penetrates into the solid, the ions undergo continuous small angle forward scattering through interactions with the electron density between the atomic nuclei, continuously losing small amounts of energy as a function of depth penetrated. In addition, a small fraction of the ion beam undergoes head on collision with the nuclei of the atoms present in the solid (the target atoms) and is consequently back scattered. The back scattering results in a major loss of energy (transferred to the atoms struck, which moves forward like in billiard ball collisions) from the probe ion beam, the amount of which is characteristic of the mass of the target atoms. Determination of the energy losses suffered by the back scattered ion beam after exit from the solid thus allows, in principle, identification of the atoms present, their depth distribution, and, from intensities, the atomic concentrations. The depth probed can vary from 1–20 µm depending on ion species used, ion energy, and instrumental set-up. At these high energies, the ion/material interactions are well described by classical kinematic equations, and quantification can be achieved without standards. The depth resolution also varies strongly, depending on the material, ion energy, and instrumental set-up, plus the depth probed (poorer at greater depth) ranging from a best of about 2 nm to 30 nm. Element sensitivity is a strong function of the position in the periodic table, varying as Z squared. It can be as poor as 10 atomic percent at the low Z end and as good as 10 ppm range at very high Z. In principle all Z values above that of the probe ion can be detected (so not H), but the poor energy resolution of the solid state detector used means that some elements cannot be distinguished from adjacent elements in the periodic table. Lateral resolution is limited to the beam diameter, usually 1 to 4 mm, but a few specialized microbeam systems do exist.

The main use of RBS is nondestructive element depth profiling of thin films in the few µm depth range. If the target is crystalline, specialized modes involving beam alignment with crystallographic axes can give structural information, such as the degree of lattice damage.

In Scanning Electron Microscopy (SEM), a probe electron beam (typically a few hundred eV to a few 10's keV energy) is finely focused (down to 10 Å capability in some instruments) and scanned over a solid surface. The interaction of the beam with the sample material generates a variety of responses, including fluorescence emission, which can be used for elemental analysis (see the EDS summary).

One of the major responses is a copious emission of low energy secondary electrons (0 to 50 eV range), which escape from the top few 10's Å of the material. The secondary electron yield strongly depends on the angle of impact between the probe beam and the local surface topography, so rastering the beam across the surface produces a changing intensity with changing topography. These changes in secondary electron emission intensity are used to modulate the brightness of a synchronously rastered cathode ray tube, creating an image. The image can be highly magnified up to 500KX, with a lateral resolution determined by the diameter of the probe beam, and have the look of an optical image (though the depth of focus is much greater).

In addition to the secondary electrons, a much smaller quantity of backscattered primary electrons are produced. Their intensity depends strongly on material (high for high Z elements), so the image also contains some Z contrast. Filtering out the low energy secondaries enhances this contrast (backscatter mode), producing an average Z dependant image instead of topography image.

The SEM is often the first or second (after an optical microscope) technique used to provide a magnified image of an area of the sample to be examined. It is often used in conjunction with an ancillary analytical technique, such as EDS, to provide elemental analysis capability to go with the imaging.

Scanning Transmission Electron Microscopy (STEM)* 21

Scanning Transmission Microscopy (STEM) is a specialized form of TEM (see TEM summary) where the high energy electron beam (100 to 300 keV) is focused to a small spot (down to a minimum of about 3 Å) and scanned across a sample which is usually a material which has been sectioned to a very thin film (5 to 500 nm thick). After interaction with the material of the film, the transmitted beam can be subjected to various treatments, as in TEM (bright and dark field imaging; convergent beam electron diffraction, CBED; and electron energy loss spectroscopy, EELS). Also, the emitted X-ray fluorescence can be analyzed by EDS (see EDS summary). Keeping the sample to a very thin section stops the lateral spread of the primary beam as it passes through, allowing the spatial resolution to be determined by the diameter of the focus beam, down to almost atomic dimensions.

The major use of STEM is for imaging and compositional analysis (sometimes also chemical state analysis, from EELS) at very high spatial resolution across thin film material interfaces, such as found in semi-conductor and other high technology devices. Single atom imaging is possible for heavy atoms. The disadvantage is that a very thin interface cross-section must be first prepared. Also, at very high resolution, the electron beam is focused into such a small area that the local electron density becomes extremely high and can anneal, damage, or even burn through the film if the dwell time is too long.

In Scanning Tunneling Microscopy (STM), a sharp tip is brought to within a few Å of a conducting surface in ultra high vacuum and held at a positive or negative potential with respect to the tip. At this distance a quantum mechanical electron tunneling current flows between an individual atom on the surface and an atom on the tip (or vice versa, depending on the polarity). The magnitude of the current is an exponential function between the two atoms involved. So scanning the tip across the surface therefore produces a strongly varying current as the topography and distance between tip and surface atoms changes. This allows mapping of the topography (actually the surface electron density) with a resolution of 0.01 Å in the Z direction, and atomic resolution in the XY direction.

In the related technique, Scanning Force Microscopy (SFM; also known as atomic force microscopy, AFM), the tip is mounted on a cantilever and the Van De Waals forces between the surface and the tip deflects the cantilever. The deflection is also a strong function of the separating distance, allowing topography mapping, as in STM. The resolution in SFM is about ten fold poorer than STM, but the technique has the great practical advantages of being operated in air and not requiring a conducting sample.

The major use of STM is in the research area of imaging atomic structures on clean conducting and semi-conducting surfaces, under UHV conditions. SFM, on the other hand, is widely used throughout industry as a very high resolution surface profilometer to monitor surface roughness, defects, and man-made micro or nano structures.

WALTER JOHNSON and CHUCK YARLING

Sheet resistance measurements are a way of electrically characterizing semiconductor wafers, thin films on wafers, and thin near-surface regions of diffused or ion-implantation doped wafers. The method of choice for the resistance measurement is a four point probe arrangement of contacts with the surface through which current is passed using two of the probes and the voltage drop being measured across the other two.

Parameter measured	Electrical resistivity
Main usage	Quality control of Si wafers at various processing stages
Sample requirements	Flat surface, a few square millimeters in area
Destructive	No, unless depth profiling by beveling sample
Accuracy	1%
Quantitative	Yes, but calibration needed
Lateral resolution	mm usually
Depth resolution	None, except by beveling sample
Instrument cost	$10,000–$100,000
Size	Benchtop

Introduction

The four point probe is the most common tool used to measure sheet resistance. The normal range for sheet resistance in semiconductor processing is from less than $0.02\ \Omega$/square for aluminum films to about 1 MΩ/square for low dose implants into silicon.

Until the advent of the dual configuration technique,[1] the construction of a high precision tester,[38] and control over contact resistance, accurate sheet resistance measurements were usually only possible with van der Pauw structures. These advances brought the consistency to a level where the four point probe could be used for equipment and materials characterization. Sheet resistance is a widely used method for measuring the electrical properties of films in the semiconductor industry. Typical applications include starting silicon substrates, ion implantation, diffusion, epitaxial silicon, polycrystalline silicon, metals and silicides.

Principles and Definitions

Materials can be classified by their ability to conduct electricity[3]: insulator, conductor, and semiconductor. Insulators cannot be measured with the four point probe and often are the source of problems in measuring conductive or semi-conductive films.

Metals tend to give up valence electrons freely for conduction, resulting in all the electrons bonding all the atoms in an electron sea.[4] Electrical resistivity results from scattering of these electrons by the lattice.[5]

Silicon is an insulator at a temperature of absolute zero (0 K). There are no free carriers to conduct current. As temperature rises within the bulk material, bonds within the silicon lattice structure begin to break up, providing the free carriers necessary for current flow. Both n, the number of electrons per cubic centimeter, and p, the number of holes per cubic centimeter, depend on the temperature of the material. Since each broken bond produces both an electron and a hole, both n and p are present in equal numbers in pure silicon. Thus, at room temperature ($T = 27\ °C = 300\ K$),

$$p = n = n_i = 1.45 \times 10^{10}/\text{cm}^3$$

Donors and Acceptors

As previously stated, intrinsic silicon has an equal number of donors, n, and acceptors, p, the number of which is dependent upon the ambient temperature. However, it is the interaction of adjacent semiconducting layers having different densities of dopants that give semiconductor materials their unique properties. These adjacent layers are usually, but not necessarily, doped with carriers of opposite dopant types. Doping may take place by several methods, including diffusion, implantation, chemical-vapor deposition (CVD), and spin-on glass (SOG).

Dopants are impurity atoms added to the semiconductor silicon during crystal growth or subsequent processing to enhance its ability to carry current (its "conductivity").[3, 6] For example, silicon is doped with arsenic or phosphorous (and sometimes antimony) to produce an n-type material. For p-type materials, silicon is normally doped with boron. The total number of donor, or n-type, dopant atoms is N_d. The total number of acceptor, or p-type, dopant atoms added to silicon is Na

Mobility

Mobility, or drift mobility, μ, is strongly dependent upon process conditions (such as anneal parameters) for high dopant conditions.[3, 6] Mobility is defined as the "ease" with which the carriers move through bulk material. Conditions within the bulk material that may affect this property include carrier concentration, temperature, crystal defects, presence of undesired impurities, and other scattering mechanisms.

Mobility is quantitatively defined by the following formula:

$$\mu = D \cdot (q/kT), \quad \text{cm}^2/\text{V·s}$$

where D is the diffusion constant, a bulk property of silicon, in cm^2/s; q is the electronic charge; k is Boltzmann's constant (1.38×10^{-23} J/K); and T is temperature in kelvin.

When the dopant type and total impurity concentration of a silicon sample are known, the mobility and diffusion coefficient may be obtained from calibration curves relating mobility to dopant concentration.[3, 6]

Conductivity

Once impurities have been added to the silicon, the material is now "quasi-" or "semi-" conducting. Conductivity, σ, is the ability of the silicon to conduct electron and hole flow.[3, 6] Conductivity is defined:

$$\sigma = (qn\mu_n) + (qp\mu_p), \quad (\Omega \cdot cm)^{-1}$$

where q is the charge of an electron (1.6×10^{-19} coulombs); n is the number of electrons in carriers/cm^3; p is the number of holes in carriers/cm^3; μ_n is the electron mobility in cm^2/V·s; and μ_p is the hole mobility in cm2/V·s.

In most cases, one of the dopant types, either donor (N_D) or acceptor (N_A), far exceeds the number of the other. When this happens, this equation is reduced to only one term. The equation then becomes

$$\sigma = (qn\mu_n), \quad (\Omega \cdot cm)^{-1} \quad \text{if } N_D \gg N_A$$

or

$$\sigma = (qp\mu_p), \quad (\Omega \cdot cm)^{-1} \quad \text{if } N_A \gg N_D$$

Resistivity

The resistivity of a material[7, 8], is a measure of its ability to oppose electrical conduction that is induced by an electrical field placed across its boundaries. Resistivity, p, is defined as

$$\rho = 1/\sigma = \sigma^{-1}, \quad \Omega \cdot cm$$

The resistivity of the three classifications of materials used in semiconductors is as follows:

insulator	$\rho > 10^3 \; \Omega \cdot cm$
semiconductor	$10^{-3} < \rho < 10^3 \; \Omega \cdot cm$
conductor	$\rho < 10^{-3} \; \Omega \cdot cm$

Resistivity also varies with temperature. Temperature coefficients of resistance (TCR) can be found in the literature.[9] Pure metals commonly used in semiconductor manufacturing have TCRs in the range of +0.2 to +0.4%/°C.[10] There are several cases where negative temperature coefficients of resistance can be obtained for very thin, low density or cermet films.[5] Doped polysilicon layers can also exhibit negative TCRs.

Resistance

Once the resistivity of a silicon sample is known, its resistance may be determined. Resistance, R, is defined as

$$R = \frac{\rho L}{A} = \frac{\rho L}{tW}, \quad \Omega$$

where ρ is the resistivity in $\Omega\cdot$cm; L is the length of the sample in cm; and A is the cross-sectional area of the sample being probed in cm^2 = thickness (cm) $\times$ width (cm) = t (cm) $\times$ W(cm).

Sheet Resistance

Free-standing thin layers can be usefully described with the term sheet resistance, R_s, rather than by the bulk resistance. This is more descriptive, since it involves the thickness, t, of a layer. Sheet resistance[11] is calculated from the resistance of the sample by using the formula above, which for $W = L$ reduces to

$$R_s = \frac{\rho \ (\Omega\cdot\text{cm})}{t \ (\text{cm})} \quad \text{in } \Omega, \text{ but more commonly referred to as } \Omega/\square$$

The value of the concept of sheet resistance lies in the ability to calculate the resistance of large areas knowing only the size of the sample and its resistivity. For instance, if a bar of silicon is twice the length of another bar (of the same resistivity), the resistance of the longer bar is twice that of the shorter one, but the sheet resistance, R_s, remains the same.

The Four Point Probe Method of Measuring Sheet Resistance

Probe assemblies for the four point probe historically have probe tips which force a current and measure the resulting voltage drop, for the sheet resistance measurement, in one of two arrangements: linear or square arrays. Since the linear probe-tip array that is perpendicular to the edge of the wafer is the most prevalent, only this arrangement will be considered in this discussion.

Sheet resistance is measured by passing a current between two probe pins and measuring the resultant voltage drop across two other pins. In the linear array, it is customary for the outer pins to carry the current and the inner two to measure the voltage. Geometric effects that interfere with the sheet resistance measurement, such as the probe's proximity to the edge of the wafer (which causes "current crowding"), can be accounted for by a technique called dual configuration. This technique requires that two resistance measurements be made.

The first measurement, R_A, is obtained in the traditional manner, as shown in Figure 1a, with the outer pins carrying the current and the inner pins measuring the voltage. The second measurement, R_B, is obtained by having the first and third pins carry the current and using pins two and four to measure the voltage, as shown

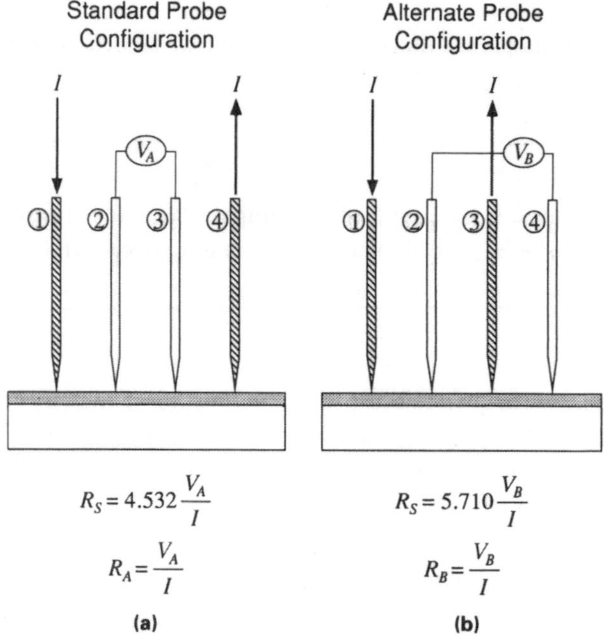

Figure 1   **Paired measurements which are used to calculate sheet resistance in the dual configuration mode.**[1]

in Figure 1*b*. The use of both arrangements allows for the correction of discrepancies in pin spacing and any current crowding effects.[1]

Measurement Considerations

Selecting the optimum probe head and measurement current for some samples can be a difficult task. Usually we settle for some acceptable set of conditions that are determined with the minimum of effort. Metal films seldom exhibit measurement problems due to the good ohmic contact created in the metal pin to metal film contact. For some samples we need to exercise a little more care in investigating the conditions on the wafer and the type of probe used to find conditions that give the most accurate and repeatable measurements. For example, in the case of an implanted layer, a few things to consider are dose, energy, species, background concentration, screen oxide, and surface preparation. There are a number of combinations that create conflicting measurement requirements, thereby necessitating compromise.

Requirements for an ideal measurement:
- minimum contact resistance
- maximum surface concentration
- minimum leakage across the surface
- minimum leakage across any layer boundary or junction.

If all the above requirements are met, then there is a good chance of getting excellent results. Some ways to achieve these requirements are discussed below.

Minimum contact resistance is obtained by ensuring that any unwanted insulator layer, such as a thick oxide, is removed from the surface; deposited oxides will cause measurement problems if not effectively removed prior to measurement. Native oxides will be present on most layers, but in most cases the standard 100 mg loading probe head will effectively puncture this native oxide to make good contact at the conductive layer. This does rely, however, on the probe head having some surface roughness or on its being sufficiently sharp. An inversion layer common on p-type silicon substrates[12, 13] usually can be bypassed by using a sharp probe with high gram-loading that can break through this layer to make contact with the more conductive layer, the one we wish to characterize. More current can sometimes help break down this layer.

High contact resistance can result from contaminated probes, particularly if the probes are used on metallic layers and the metal has transferred to the probe tip.

Another cause of high contact resistance is low surface concentration of dopant in diffused or implanted layers. This effect will normally manifest itself in poor probe qualifications; it can also give rise to random bad site measurements or nonreproducible maps. If the high-resistance surface is part of the layer measured, then increasing the surface area of the four-point probe pins that contact the sample will decrease the total contact resistance (i.e., use a larger probe tip radius).

Maximum surface concentration can be obtained for implant monitors by reducing the background concentration (i.e., using a higher resistivity substrate). This increases the resultant concentration after the implant. A screen oxide can also be used, capturing the low-concentration tail into the oxide layer. This tail is consequently removed upon removal of the oxide.[1]

If low surface concentration seems to be the problem, the first requirement is to better understand the physical characteristics of the layer being measured. Parameters such as species, dose, energy, background concentration, screen oxide, and surface preparation should be considered, as well as the use of simulations predicting the likelihood of success. In some cases sharper probes or higher loadings may be recommended; in other cases, different surface treatments may be tried or alternative substrates recommended for the test wafers. In all cases, the objective is to be sure to characterize the material over the depth of interest, not a surface region which may vary but is of no device performance interest (e.g., native oxide or a substrate material).

Native oxides incorporate a number of various contaminants and can often cause poor sheet resistance measurements, especially on silicon where the resistivity is relatively high. The surface leakage is reduced by ensuring a clean, fresh surface with a very thin, stable SiO_2 layer on silicon. Typically this involves an HF dip, rinse, H_2SO_4–H_2O_2, and rinse procedure for silicon wafers.

The maximum substrate junction resistance would, of course, be obtained by putting the conductive layer on an insulator. This is not always possible or practical. For implants, the only real alternative is to use a monitor wafer of opposite conductivity type with a lower background concentration. An epi junction width can be narrowed with a high resistivity substrate or a process change. High resistivity substrates will help, where metal layers must be deposited directly on silicon.

Troubleshooting

Some problems commonly encountered making sheet resistance measurements are:
* oxidizing metal layers
* punch-through of thin layers by the probes to other layers or the substrate
* wear of probe tips, making them smoother and rounder
* contamination of probe tips, increasing their contact resistance
* measurement current too high, causing layer heating and an apparent rise in sheet resistance
* measurement current too high, causing substrate leakage and an apparent fall in sheet resistance
* substrate resistivity too low for the implanted dose, causing substrate leakage.

A good starting point in any troubleshooting procedure is to refer to a known standard. In this case, it is desirable to set aside test wafers (or "golden wafers") which have undergone the same process as those being currently tested and which have given good measurement results in the past. Re-measurement of the test wafer will now give an immediate sign of whether the problem lies in the wafer or the probes. Although troubleshooting is possible without such wafers, the task becomes more time-consuming.

When trying to make measurements on shallow implants and diffusions or thin metal layers, it may be possible to "punch-through" the layer when the probes used are too sharp. This allows leakage into the substrate, effectively providing current paths in parallel and giving apparently lower sheet resistances. The remedy is to use a flatter probe or higher resistivity substrates. In some circumstances we need to compromise on a set of conditions that gives, as before, reproducible measurements, as shown by good probe qualifications.

After extended use, probe tips become worn, and even after probe conditioning, the surface irregularities that break through the native oxide will no longer be present. It will then be difficult or impossible to get acceptable results. When this occurs, the probe should be replaced.

Probe tips will inevitably pick up contamination after use, either from the wafers they probe, or from handling, or from the air. The usual result of probe contamination is an increase in the number of bad sites during mapping and poor probe qualifications. To remove lightly bound contamination, brush the tips lightly with a fine brush or a Q-tip soaked in a fab solvent such as IPA. As usual, a probe qualification will reveal the effectiveness of this treatment. Tightly bound contamination may require a probe conditioning, the necessity of which will vary with each application.

Probe Qualification

One of the main sources of noise in the sheet resistance measurement is the probe itself. To get accurate and repeatable results, the probe head pins must make good electrical contact each time they touch the surface of the wafer. The probe qualification procedure should check the short-term repeatability of the probe head.

For example, if a probe used to measure a process with a uniformity specification of 0.75% has a repeatability (noise) of 0.5% and the true wafer uniformity is 0.65%, then the net result is

$$S = (S_{\text{wfr}}^2 + S_{\text{probe}}^2)^{1/2}$$

where S is the resulting standard deviation, S_{wfr} is the real standard deviation of the wafer, and S_{probe} is the probe qualification standard deviation.

The monitor standard deviation of 0.82% is larger than the uniformity specification and therefore out of specification. From this example we may conclude that the tighter the process monitor specification, the smaller should be the allowable noise introduced by the probe head. As the probe repeatability degrades, the quality of the map degrades first. The second parameter to degrade is the standard deviation; the last is the average sheet resistance. To assure the continued quality of the measurements and maps, the probe repeatability should be checked regularly. It is also essential to use the right probe head for each application. Probe repeatability is best checked on monitor wafers specially chosen for this purpose. If the repeatability is being checked on ion implanted wafers, a wafer should be used that has a resistivity equal to or greater than that of the wafers typically measured. Depending on the process monitored, it is essential to keep a few monitor wafers set aside. These monitor wafers should be changed about every three months, depending on their type and usage and on the type of probe head used.

Conclusion

Many factors govern the success or failure of sheet resistance measurements. Methodology exists for troubleshooting poor or doubtful sheet resistance results. It is useful to consider the following generalized statements when preparing to measure a wafer or investigating problems:

- When investigating suspect results, perform a probe qualification or repeatability on the suspect wafer. If the probe qualification is good, then the results are believable. If the probe qualification is bad, carry out a second probe qualification on a monitor wafer with similar process history and sheet resistance. If these results are good, then the wafer has a problem; if bad, then the probe has a problem.
- A good probe qualification usually identifies a probe that will give a meaningful map with explainable features and accurate sheet resistance.
- A map with inexplicable rejects or erratic high and low points usually means a measurement problem and erroneous standard deviation.
- If there is a contact resistance problem, a poor probe qualification will result.

This summary relates closely to Technique Summary 24, Spreading Resistance Analysis (SRA). SRA refers specifically to a resistivity, or sheet resistance, profile made along a beveled surface cut from the sample; that is, it measures resistivity as a function of sample depth. The use of two point probes for making this measurement is described in the SRA summary. Four point probes can also be used and, for flat surfaces, is the preferred method.

References

1   D. S. Perloff, J. N. Gan, and F. E. Wahl. "Dose Accuracy and Doping Uniformity of Ion Implantation Equipment." *Solid State Tech.* **24** (2), 1981.

2   W. A. Keenan, W. H. Johnson, and A. K. Smith. "Advances in Sheet Resistance Measurements for Ion Implant Monitoring." *Solid State Tech.* **28** (6), 1985.

3   P. E. Gise and R. Blanchard. *Semiconductor and Integrated Circuit Fabrication Techniques.* Reston Publishing Co., Reston, VA, 1979

4   M. J. Starfield and A. M. Shrager. *Introductory Materials Science.* McGraw-Hill, New York, 1972.

5   L. I. Maissel and R. Glang. *Handbook of Thin Film Technology.* McGraw-Hill, 1983.

6   R. S. Muller and T. I. Kamins. *Device Electronics for Integrated Circuits.* Wiley, New York, 1977.

7   J. C. Irvin. "Resistivity of Bulk Silicon and of Diffused Layers in Silicon." *Bell System Technical Journal.* **41** (2), 387–401, 1962.

8   E. Hesse. "Resistivity Measurements of Thin Doped Semiconductor Layers by Means of Four-Point Contacts Arbitrarily Spaced on a Circumference of Arbitrary Radius." *Solid State Electronics.* **21**, 637, 1978.

9   "Standard Method for Measuring Resistivity of Silicon Slices with a Collinear Four-Probe Array." *Annual Book of ASTM Standards.* Vol. 10.05, ASTM F84, ASTM, Philadelphia, 1988.

10   R. C. Eeast and M. J. Astle. *CRC Handbook of Chemistry and Physics.* 63rd ed., CRC Press, Boca Raton, FL, 1982–83, p. E-81.

11   F. M. Smits. "Measurement of Sheet Resistivities with the Four-Point Probe." *Bell System Technical Journal.* 711–718, May 1958.

12   S. J. Pearton. "Near-Surface Modifications Induced by Hydrogen in Semiconductors." *Proceedings.* Second International Conference on Solid State and Integrated Circuit Technology, Nov. 1989.

13   P. Kramer and L. J. Van Ruyven. "Space Charge Influence on Resistivity Measurements." *Solid State Electronics.* **20**, 1011–1019, 1977.

14   W. A. Keenan, W. H. Johnson, and A. A. Smith. "Advances in Sheet Resistance Measurements for Ion Implant Monitoring." *Solid State Tech.* June 1985.

ROGER BRENNAN and DAVID DICKEY

Spreading resistance analysis (SRA) is a method of dopant profiling that relies on the relationship between dopant concentration and electrical resistivity in semiconductors. In SRA, the resistivity profile is determined from resistance measurements made between two probes at a series of locations along a beveled surface of the sample. The method is used extensively on silicon, where it is generally an inexpensive way of obtaining a wealth of information. The accuracy for dopant concentrations is generally in the 10–30% range; and that for depth is typically better than 3%. The basic spatial resolution is on the order of microns, while depth resolution can approach 10 Å. SRA is sensitive to the full range of resistivity values found in both n- and p-type silicon. It is useful with a wide range of other materials, although the requirement for calibration seriously limits the accuracy in many cases.

Parameter measured	Electrical resistivity
Main usage	Determination of laterally resolved dopant concentrations for semiconductors, as a function of depth
Sample requirements	Surface of at least 20×20 µm area
Destructive	No, except for sample beveling
Detection limits	Carrier concentrations down to 10^{11} cm^{-3}
Quantitative	Yes, but calibration needed
Lateral resolution	A few microns
Depth resolution	~10 Å
Depth profiling	Yes, but requires beveling the sample
Instrument cost	Variable, but not less than $10,000
Size	Benchtop

Basic Principles

SRA determines dopant distribution by measuring the electrical resistivity of the silicon rather than counting atoms. Fortunately, resistivity is connected with dopant concentration in a semiconductor:

$$1/\rho = \mu q N \tag{1}$$

where ρ is the resistivity in $\Omega\cdot$cm; μ is the carrier mobility in cm^2/V·s (V·s = $\Omega\cdot$C, which can be obtained by substituting A = C/s into Ohm's law); N is the dopant concentration in cm^{-3}; and q is the charge of an electron (1.6021×10^{-19} C).

Measuring electrical resistivity therefore determines dopant concentrations. The spreading resistance probe has high spatial resolution and can detect variations in dopant concentration over depths as small as a few tens of angstroms. A test voltage is introduced into the specimen and the resistance is measured. Using the appropriate equations, one can extract values of resistivity from these simple measurements.

Methodology

Two probe tips, usually osmium, are mounted on the end of separate arms. Each arm pivots on a kinetic bearing system that virtually prevents lateral motion of the probe tip. The probe tips are positioned very closely together, less than 20 μm apart. The probe tips are lowered onto the sample as gently as possible, because of the high pressure of the probe tip—over a million pounds per square inch. The probe tip is hard enough to fracture the silicon, leaving "probe marks."

Typically, 5 mV is applied across the probes, and the spreading resistance is measured at a series of depths (see Figure 2). If the contact area of the probes is made small enough, the measured resistance is mostly from the current crowding in the immediate vicinity of the probe tip. This is determined by:

$$R_S = \rho / 2a \tag{2}$$

where R_S is the measured spreading resistance in Ω; ρ is the local resistivity in Ω·cm; and a is the radius of the contact area of the probe in cm.

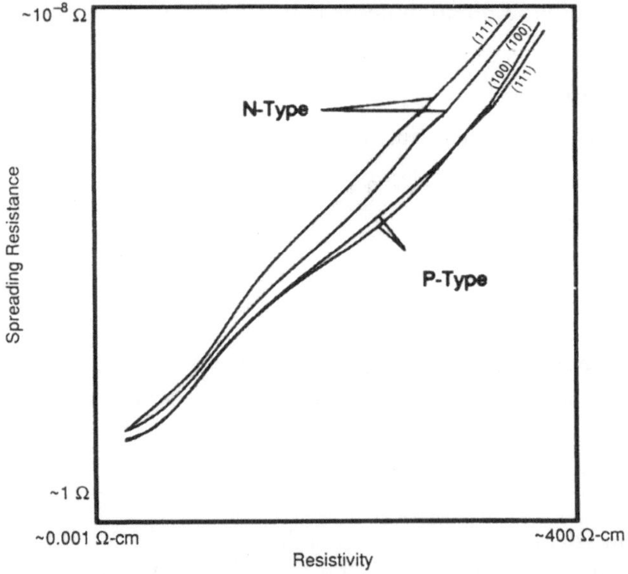

Figure 1   **Calibration chart of spreading resistance versus resistivity.**

This expression does not reflect most situations. Usually, calibration bulk samples with well-documented resistivity are measured and a is then calculated. The contact areas of the probes are seldom truly circular or the same size. When unconstrained, the resistivity is sensed in a sampling volume that approximates a hemisphere with radius a. Corrections are needed if there is an appreciable resistivity gradient through the sampling volume or if the sampling volume is distorted due to the proximity of a p–n junction, an insulator, or a region of much higher conductivity, such as a buried layer. Several data reduction procedures have been developed for these corrections.

An extensive calibration is usually required because of deviations from Equation 2. One calibration method would be to use 64 pieces with well-documented resistivities divided into four groups: p<111>, p<100>, n<111>, and n<100>. Each group has 16 samples ranging from 0.001 to 400 Ω·cm. In Figure 1 the calibration curves show departures from the straight line relationship of Equation 2. The n-type material produces higher spreading resistance readings than does the p-type of the same resistivity, and no material of a given resistivity produces higher spreading resistance readings than does n<100>. At low resistivities, the slope of the calibration curve becomes flatter as the resistivity decreases. At the high end of the calibration curve, the slopes are relatively constant and can be used to estimate resistivities greater than 400 Ω·cm.

Because the spreading resistance probe senses the resistivity in the sample direcdy under the probe tip, one can angle lap a silicon structure, probe down the beveled surface (see Figure 2), and obtain a resistivity-versus-depth profile. A plot of carrier concentration versus depth can be calculated using published mobility values.[1] Plots of carrier concentration versus depth are more useful because they tend to follow the doping profile. The spreading resistance probe does not directly sense the doping type. An additional test involves heating one of the probe tips and determining the polarity of the Seebeck voltage to find out conductivity type.

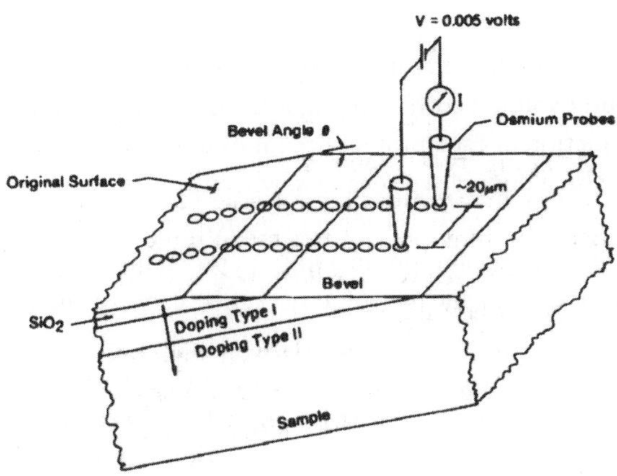

Figure 2   **A spreading resistance measurement in a beveled sample.**

Sample Preparation

Sample preparation is important. The sample must be beveled to the appropriate depth, and care must be used to avoid rounding the beveled edge, which would compromise the depth accuracy. Patterned samples should not be probed any closer to the diffusion mask edge than half the probe spacing. Also, the sample should not be beveled closer than half the probe spacing to the edge of the diffusion mask. Samples should be beveled immediately before probing. Long waiting periods, such as overnight, tend to produce noisier data. The lapping method should be totally mechanical to minimize nonstable surface charges.[2] The resultant bevel angle should be measured. Scratches on the beveled surface should be minimized because they change the probe contact area and thus compromise the calibration. Usually ion implants should be annealed prior to SRA.

Probing

Spreading resistance measurements are affected by noise, such as scatter in measured resistance on known homogenous material. The probe tips contact the silicon in different ways because the probes are lightly loaded and the contact area is very small. This causes differences in the measured resistance, but the noise can be reduced by increasing the load on the probe tip. Increasing the load increases the sampling volume and reduces the depth resolution. Typically, a probe tip loading of 2.5 g is used for shallow structures having junction depths less than 0.5 µm and a loading of perhaps 10 g for structures having junction depths 1 µm or deeper.

The probes must travel in a path relatively free of scratches and perpendicular to the bevel edge; the step size must be sufficient to prevent the probes from stepping into the previous probe marks. A well-prepared pair of probe tips should have low noise, a small sampling volume, minimum penetration, close spacing, and physical stability and compact size and should cause minimal damage to the silicon.

Data Reduction

The spreading resistance measurements are loaded directly into a computer containing the calibration data and the sampling volume correction algorithms. However, much human input is needed for data reduction of each profile run, for example, bevel edge, start point, doping types, points at which doping type changes, when to apply sampling volume correction, crystal orientation, step size, bevel angle measurement, adjustment for stray points, and measurement of probe separation. In addition, problems may arise with very shallow, very low dose surface layers due to geometrical effects and/or surface charges.[3]

When differences in resistivity are relatively small, the relatively high noise of the spreading resistance measurement makes data reduction uncertain. Data reduction is difficult at low resistivities due to the flatness of the calibration curve in this region. At concentrations above 10^{19} cm^{-3}, SIMS can be more accurate than SRA, but SIMS senses atomic concentration and SRA senses carrier concentration. There can be a large difference between these measurements at low resistivities; however, sometimes a combination of SRA and SIMS is needed.

Applications

Typical applications of SRA include predepositions and drive-ins, epi thickness and profiles, full bipolar transistor structures, starting material, buried layer-up diffusion, poly resistivity and thickness, autodoping, and ion implant doses and depths. SRA is useful in checking the quality of incoming material, developing and documenting a new process, modifying a process and confirming the results, checking the quality of production processes, and finding out why the wafer sort yield went up or down.

Dedicated Spreading Resistance Test Patterns

To do SRA, one needs to identify a suitably large area containing the desired structures. The area needed—one of the smallest areas required by any profiling technique—can usually be found on the chip, but this is often less than optimal. As a long-term solution, dedicated spreading resistance test structures are suggested (see Figure 3). Sooner or later, a new test pattern will be laid out anyway, and often these features can be included with negligible time and expense. The spreading resistance test patterns offer minimal analysis cost, complete characterization of the wafer fab process, optimum resolution, and clear, straightforward identification of all possible doping structures. By selectively exposing the silicon at the appropriate masking steps, one can produce rectangles that contain every structure of possible interest.

Comparison with Other Techniques

SRA Versus SIMS

To characterize, for example, the deep-diffused power transistor referred to in Figure 4 requires seven decades of dynamic range; SIMS is considerably more limited (five decades). The power device needed to be profiled down to 120 µm; this is too deep for SIMS to depth profile economically. SRA detected a high resistivity region from about 40 to 60 µm; few, if any, other techniques can detect this.

SRA Versus MOS C–V and SIMS

Analyzing the blocking ability of a material (such as photoresist, oxide, or nitride) to prevent dopant from getting into the silicon requires checking four possibilities: (1) no dopant penetration into the silicon, (2) some dopant penetration, (3) more dopant penetration, and (4) considerable dopant penetration. SRA is the only way to catch all these possibilities in one pass. MOS C–V would characterize (1) and (2) very well, but would have trouble with (3) and serious difficulties with (4); SIMS would be helpful on (4) only.

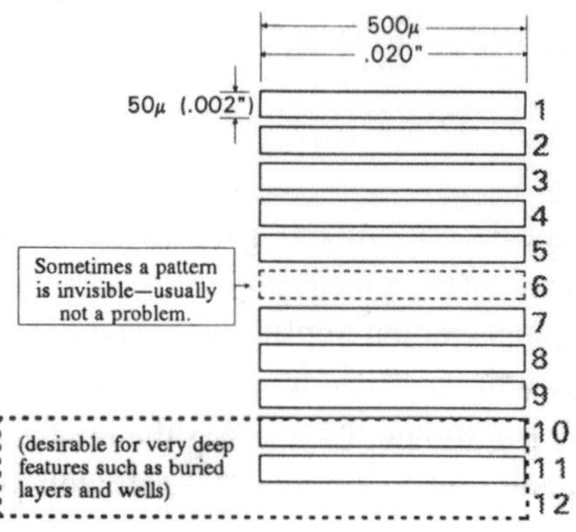

NOTES

1. If there is a very shallow emitter/base and a very deep buried layer a 100μ x 1000μ structure is suggested.
2. In general, longer is always better—say up to 1500μ.
3. Width can be a problem as the number of patterns increases. Please reduce the space between patterns to what is essential for lateral diffusion effects.
4. Label the structures at the first possible masking step.
5. Orient the long dimension of the pattern parallel to the major flat.. This is strongly recommended for epi on <111> substrates.
6. Make one common buried layer when possible. (Also wells.)
7. In relatively high beta (narrow base) bipolar devices the base width should be characterized with and without the buried layer underneath it.
8. Top side passivation is an annoyance for us (causes scratches and slows down beveling.) Please *consider* etching the entire region at pad mask. Perhaps some etching tests should be made first.
9. Sometimes allowances for "invisible" patterns, large epi shifts, and huge silicon step heights need to be made.
10. One or more back-up sets would be appreciated.
11. After dice separation, you may wish to keep the test dice as part of the permanent record for that lot of wafers.
12. If all dice must be product dice and you are exposing with a stepper, you could place these structures two high in the scribe lines. It is better than nothing.

This list of notes is surely incomplete. Although SRA was developed in the early sixties, the technique continues to evolve (i.e. there are still more subtleties to identify, resolve and document).

Figure 3   **Illustration of dedicated spreading resistance test patterns.**

Limitations

Only the net *n*- or *p*-type concentration of the electrically active dopants are sensed by SRA. SRA is destructive and highly technique-dependent and can have substantial uncertainty in resistivity measurements. C–V analysis may be more accurate for

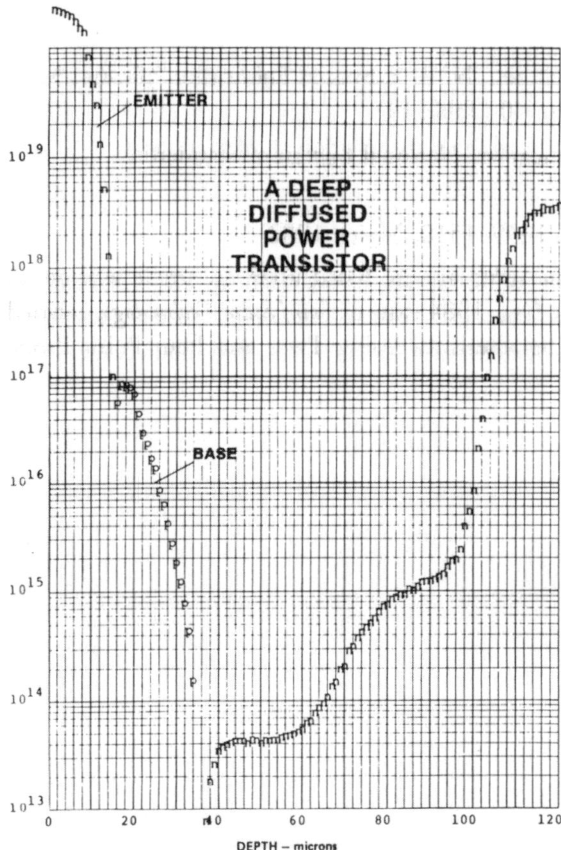

Figure 4   A deep-diffused power transistor

channel regions in MOS devices, and SIMS may be more accurate for concentrations greater than 10^{19} cm^{-3}. In addition, problems have been experienced with very shallow, very low dose surface layers due to geometrical effects and/or surface charges. Equipment maintenance and calibration are relatively involved; without professional dedication, many serious mistakes can be made.

Conclusions

The dopant distribution controls almost all electrical properties of semiconductor devices. To produce a semiconductor device with the desired electrical properties, the dopant distribution must be known. SRA monitors dopant distribution over more than eight decades; no other single analytical technique can match this range. SRA also provides general purpose, and relatively inexpensive, depth profiling with high spatial resolution and excellent depth accuracy and with unmatched sensitivity at low doping levels.

References

1   Thurber, Mattis, Liu, and Filliben. N. B. S. Special Publication 400–64, May 1981, p. 34 (Table 10) and p. 40 (Table 14).

2   J. Ehrstein. Private communication. National Bureau of Standards, Gaithersburg, MD.

3   S. M. Hu. *J. Appl. Phys.* **53**, 1499–1510, March 1982.

Several sections of this technique summary are based upon an article written by R. Brennan and D. Dickey in the Dec. 1984 issue of *Solid State Technology*, entitled "Determination of Diffusion Characteristics Using Two- and Four-Point Probe Measurements."

In Static Secondary Ion Mass Spectrometry (Static SIMS), a primary beam of ions (typically 1 to 10 keV) penetrates into a solid surface. The energy transferred to the lattice causes atoms, or clusters of atoms, to eventually be ejected from the surface region (from the top 1 to 3 atomic layers) as either ions or neutrals. These are mass analyzed in a mass spectrometer (either quadrupole or time-of-flight, TOF), allowing their identification. The difference, compared to dynamic SIMS (see the Dynamic SIMS summary), is that the dose of incidence primary ions is kept low (less than 5 times 1012 atoms/cm^2) so that, during analysis, the chemical integrity of the surface is maintained. Essentially every incoming ion strikes a fresh area not previously impacted. The objective is to deduce the chemistry of the outer layers by interpretation of the fragment ion pattern observed (cf. classical organic mass spectrometry).

The mass range accessible is up to 1000 amu with a quadrupole and 10,000 amu with TOF. TOF also has high enough mass resolution so that there is no ambiguity in atomic, or cluster identification (because atom isotopes have fractional masses, allowing any combination of masses to be distinguished with high enough mass resolution). The spatial resolution is controlled by the focus and energy of the ion beam, and is typically a few tens of μm, but sub-μm can be achieved in some cases.

The major use is in identification of the molecular species present at surfaces, particularly for polymers and organics. This may be for components of bulk materials (e.g., in polymer blends), or for monitoring changes caused by surface treatment, or for detection of contamination. Trace levels can sometimes be reached, but quantification is difficult and requires standards. With the high sensitivity and resolution of the TOF, the technique can also be used for detecting and quantifying trace heavy metals at surfaces.

Surface roughness can be measured by a variety of scanning profilometer methods, which monitor individual features, or by methods which give averaged information. For the profilometers there is a wide variety of methods possible: mechanical, optical, scanning electron microscope (see the SEM summary), and the scanning force microscope and scanning tunneling microscope (see the STM and SFM summaries). These instruments have a large range of capability in terms of depth resolution, maximum depths measurable, and lateral resolution (roughness frequency range). The mechanical profilometers obviously contact the surface, and may damage it. The depth resolution can be down to 5 nm, and the lateral resolution 100 nm (depends on tip radius), for the most sophisticated (and expensive). Optical profilometers can measure down to 0.1 nm depth and 300 nm lateral resolution (depends on wavelength).

SEM is useful for observing very short range roughness (i.e., sharp spikes or dips) but ineffective for long range small amplitude roughness. SFM has depth resolution down to 0.01 nm and a normal lateral resolution of about 0.5 nm, which is more than needed in most practical applications. STM can do an order of magnitude better, but only for conductors under ideal conditions in UHV environments. It is basically a research tool.

Optical scatterometry (see the summary for this technique) gives averaged, not individual, information. The depth resolution achievable is 0.1 nm RMS, and long and short range roughness can be separated (i.e., measurement provides information as a function of roughness frequency range).

Total Reflection X-Ray Fluorescence (TXRF) is a technique specially developed to detect trace contamination elements at wafer surfaces. It detects and quantifies their presence from their characteristic X-Ray Fluorescence, as in XRF (see XRF summary). The difference from normal XRF is that the X-ray beam strikes the surface at a very grazing angle (a few mrad) below the critical angle for total reflection, so that the beam never penetrates far into the material. The depth probed is only a few nm and depends on the surface roughness. In order to work, the method obviously requires a very flat surface, such as a Si wafer, and the grazing angle results in the X-ray beam intersecting a cm or more length of surface.

The main use is rapid multi-element analysis with very good trace detection limits for heavy metals (down to 109 atoms/cm^2, depending on element). Quantification is possible using standards. By varying the angle of incidence below and just up to and above the critical angle, some limited depth resolution capability is possible, including distinguishing atomic layer type contamination from 3D contamination, such as particles.

In Transmission Electron Microscopy (TEM), a very high energy monoenergetic electron beam (100 to 400 keV) passes through a thin specimen (less than 1000 nm) of diameter less than 3 mm (necessary to fit in the electron optics column). A series of post specimen lenses transmits the emerging electrons, with spatial magnification up to 1,000,000, to a detector (fluorescent screen or video camera) viewed in real time.

Any regions of the sample under the incident beam (usually a few μm diameter) exhibiting crystallinity, will diffract electrons away from the central spot, forming a diffraction pattern observable at the back focal plane of the objective lens. Like X-ray diffraction, this can provide identification of crystalline phase, orientation, and lattice parameters. In micro-diffraction, the incident beam is focused down to sub-micron areas, but this focusing degrades the diffraction pattern.

The above mode of operation is termed *Diffraction Mode*. Another mode is *Imaging Mode*. In this mode the reconstructed, highly magnified image (up to 1,000,000) is observed at the image plane of the objective lens. By placing an aperture in the diffraction plane, either the electrons in the undeflected beam (the center spot) can be imaged (Bright Field Mode), or those in a deflected beam (the diffracted or scattered electrons) can be imaged (Dark Field Mode). An amorphous sample of uniform thickness can undergo only Z contrast scattering (higher Z, more scattering), so regions of higher Z show up as dark areas on a bright background in Bright Field Mode. Any crystalline region will also show up as a dark area. In dark field, regions of crystallinity, or higher Z, will show up as bright spots on a dark background.

Yet another mode of use is High Resolution TEM (HRTEM), in which atom positions can be established by collecting electrons from both the undeflected and diffracted beams and comparing the observed phase interference patters to a simulation.

TEM, with its many modes, and often involving ancillary materials analysis capabilities such as EDS (see EDS summary), is the mainstay of material science and analysis of small volume (areas and thickness). A fully equipped TEM laboratory will have several microscopes with differing capabilities, plus all the necessary sample preparation techniques. See also Scanning TEM (STEM), where the incident beam is focused down to almost atomic dimensions and scanned across the sample.

In ellipsometry, a collimated polarized single wavelength light beam (UV, visible, or IR) is directed at the material under study, at an oblique angle, and the relative light intensity of the reflected beam is measured as a function of its polarization angle. In Variable-Angle Spectroscopic Ellipsometry (VASE), these measurements are made as a function of varying both the wavelength and the angle of incidence. For a thin film with a surface parallel to the substrate (i.e., a constant thickness) and inhomogeneities of less than about 1/10 of the wavelength being used, Maxwell's equations plus the Fresnel equations for calculating reflection coefficients at interfaces can be used to model the experimental data. The variable parameters in the modeling are layer thickness and the optical constants for the material in the layer. Surface and interface roughness and/or mixing and interface reaction can, in principle, be handled by treating the interfaces as additional layers with their own optical properties and thickness. The parameter values in the model are itterated until a best fit with the data is obtained. Though data is easy to obtain, this model-dependent aspect of ellipsometry analysis, with many floating parameters, can make it a difficult technique to use correctly (it can never give a unique fit), unless some of the parameters (such as the optical constant) are already well known, and can therefore be constrained.

VASE, and also single wavelength ellipsometry (less reliable) are extensively used in the wafer processing and other thin film industries, primarily to monitor thickness of single or bi-layers of dielectrics, optical coatings, semi-conductors, and magneto-optic and magnetic disk material. Depending on the optical constants, film thickness from 1 to a few 100 nm can be handled (or even microns for transparent material). Because the instrumentation is operated in air and the light can penetrate liquids, ellipsometry can also be used for in situ electrochemistry and biological interfaces.

In X-Ray Diffraction (XRD), a collimated beam of mono-chromatic X rays between 0.5 Å and 2 Å wavelength strike a sample and are diffracted by crystal planes present. Bragg's law,

$$\lambda = 2d \sin\theta$$

relates the spacing between planes, d, to the diffraction angle, 2θ, which is scanned to pick up diffraction from the different crystal planes present. The azimuthal orientation of the different beams also reveals the crystalline orientation. If the material is poly-crystalline, then diffraction rings, instead of a spot pattern, are formed (powder diffraction). Distortions or broadenings of the diffraction beams carry information on crystal strain and grain size.

Because X rays penetrate deeply (strongly Z dependent; at $\lambda = 1.54$ Å the absorption length is 1 mm for carbon, 66 μm for Si, and 4 μm for iron), XRD is intrinsically a bulk technique. Typically, however, large areas are used (several mm diameter), and there is sufficient intensity and spectral resolution using modern standard equipment to study films of thickness down to the 100's nm range for high Z material. If specialized instrumentation and geometries are used (Grazing Angle XRD, Double Crystal Diffraction XRD, synchrotron radiation source), sensitivities down to 10's nm are easily achieved, and mono-layer sensitivity is possible. Small area capabilities also exist (micro-beam XRD), but then thicker samples are required to compensate for intensity loss of the smaller areas.

The major uses of XRD are identification of crystalline phases, determination of strain, crystalline orientation and size, epitaxial relationship, and the accurate determination of atomic positions (better then in electron diffraction). Because of the strong Z dependence of X-ray scattering, light elements are difficult to deal with, particularly in the presence of heavy elements.

In X-Ray Fluorescence (XRF), the mono-energetic X-ray beam from a vacuum X-ray tube (W, Mo, Cr, and other targets are used to provide a range of energies) irradiates the sample, in air, causing excitations of core electrons of the atoms present. As these excited atoms decay back to their electronic ground states, light is emitted in the X-ray region. The specific wavelengths emitted are characteristic of the energy levels involved, and therefore of the atoms concerned. The X-ray fluorescent wavelengths are determined using a crystal diffractometer.

The depth probed is Z dependent, but, since X rays are involved both in the excitation and the emission, it is many µm, making XRF a bulk technique. A specialized grazing incidence (total reflection) adaptation exists, however, for analyzing material at the surfaces of semi-conductor wafers (see TXRF summary). Also, there is sufficient total intensity such that thin films (µm level or lower, depending on instrumentation and Z) can be analyzed provided there are no interfering elements in the substrate. Standard instruments have no significant spatial resolution (few mm), but micro-beam systems down to 10 µm do exist. The technique is applicable to all elements except H, He and Li. Energy dispersive X-ray analysis (see EDS summary) is a closely related technique.

The major uses of classical XRF, in air, are the identification of elements and the determination of composition for bulk materials. For thin films intensity/composition/thickness equations are used to determine the composition and thickness of individual layers in single or multi-layered stacks, such as used in the disk drive and semi-conductor industry.

In X-Ray Photoelectron Spectroscopy (XPS), also known as Electron Spectroscopy for Chemical Analysis (ESCA), mono-energetic soft X rays (usually Al K alpha at 1489 eV or Mg K alpha at 1256 eV) bombard a solid sample in ultra high vacuum. One of the interaction processes occurring is the ejection of photoelectrons, according to the Einstein Photoelectric Law:

$$KE = h\nu - BE$$

where $h\nu$ is the incident X-ray energy, KE is the kinetic energy of the photoelectron, and BE is the binding energy of the electron in the particular energy level of the atom concerned. Measurements of the KE's of the ejected photoelectrons directly determines the BE's, which, for core levels of atomic orbitals, are characteristic of the atoms concerned and therefore identify their presence. On a finer energy scale, small "chemical shifts" in the BE's provide chemical state identification (such as the oxidation state for a metal). The technique is applicable to all elements, except H and He, since they do not have characteristic atomic core levels.

For a solid, XPS probes 2 to 20 atomic layers deep, depending of the KE of the ejected electron, the angle w.r.t., the solid surface of detection, and the material. XPS is therefore a true surface technique. If measurement is made as a function of angle, depth distribution information is available over the depth probed. XPS is also used with sputter depth profiling to go beyond these depths. Modern laboratory XPS instruments can provide *practical* lateral resolution capability down to the 10 to 20 μm range. Specialized systems can go lower, but acquisition time becomes very long.

The particular strengths of XPS are semi-quantitative elemental analysis at surfaces without standards, quantitative analysis with standards, and chemical state determination for materials as diverse as biological to metallurgical.

Index